Shadows of Science

How to Uphold Science, Detect Pseudoscience, and Expose Antiscience in the Age of Disinformation

Kendrick Frazier

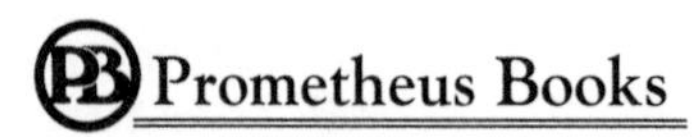

An imprint of Globe Pequot, the trade division of
The Rowman & Littlefield Publishing Group, Inc.
4501 Forbes Boulevard, Suite 200, Lanham, Maryland 20706
www.rowman.com

Distributed by NATIONAL BOOK NETWORK

British Library Cataloguing in Publication Information Available

Library of Congress Cataloging-in-Publication Data Available

ISBN 978-1-63388-938-5 (cloth : alk. paper) | ISBN 978-1-63388-939-2 (ebook)

∞™ The paper used in this publication meets the minimum requirements of American National Standard for Information Sciences—Permanence of Paper for Printed Library Materials, ANSI/NISO Z39.48-1992.

CONTENTS

FOREWORD

Richard Dawkins

Kendrick Frazier surely would have known this story, and even more surely would have been amused by it. His friend and fellow distinguished skeptic Ray Hyman was doing a double-blind trial to test the claims of a homeopath. When these claims failed, the homeopath said, with the air of one triumphantly vindicated, "You see? That is why we never do double-blind testing anymore. It never works!" Ken himself tells of a visit to China where he investigated a woman police officer who claimed she could see inside another person's body and diagnose disease. He asked her if she had ever been subjected to a controlled experiment. "She had no idea what we were talking about."

In the spirit of science, I once submitted myself to an alternative therapist who employed kinesiology as a diagnostic technique. Lying on my back I had to arm-wrestle her, I think as a test of whether "my vital energies were balanced." Or something. She then placed a closed glass vial of vitamin C on my chest and arm-wrestled me again. My apparent strength relative to hers dramatically increased. When I expressed surprise, she evidently missed my irony and gushed, "Yes, C is a marvelous vitamin, isn't it!" She was probably sincere, just massively gullible. It is to my discredit that I didn't bother to propose an experiment with a control vial. I just knew what the result would be.

Ken Frazier describes the puzzled surprise of dowsers when, under controlled experimental conditions, they failed to divine water. I had exactly the same experience when I took part in a controlled trial of British dowsers for Channel Four television. They were visibly shocked at their lack of success, and one woman actually wept. Quite simply, they had never been subjected to a double-blind trial before.

Those four stories sum up what Ken Frazier had to deal with all his life, as editor of the *Skeptical Inquirer* for four decades and leading light of the scientific skeptics movement: lamentable ignorance of the scientific method, and complete blindness to why it matters. There is only one way to discover whether a proposition about the real world is true, and that is the scientific method. If a better way were discovered, science would adopt it.

The truths that science has uncovered are breathtaking in their scope and beauty. And they keep on coming. Kendrick Frazier deeply loved science, and deep, too, was his knowledge of it. The opening and closing chapters of this book present a choice bouquet of examples, hand-picked from the garden of science. The tally is steadily increasing of Earthlike planets, basking in the goldilocks zones of their respective suns and promising the hope of alien life, a serious, scientific hope which is parsecs apart from naïve, Roswell-style ufology. Instruments sensitive to a deformity less than a proton's width have detected gravitational waves, the radiating shudder of spacetime as black holes collide at the far reaches of the universe. Proteins fold into precise three-dimensional knots whose shape is the key to the catalytic magic by which enzymes control life's processes. The 3D shape is determined by the 1D sequence of amino acids. But nobody knew how that worked. The folding of the chains just happened, and that was that: a brute fact. But now science has worked out how to predict protein folding. It may not be obvious what an astonishing achievement that is. Take my word for it.

Science, the poetry of reality, the jewel in humanity's crown, is indeed wonderful, and we who live in the twenty-first century should exult in its daily advances. But the poetry is hijacked, sometimes by cynical charlatans, sometimes by credulous fools. "This is the dawning of the Age of Aquarius." A line like that has a kind of poetic resonance, but it is fake poetry borrowed from the real romance of the stars. The garden from which Frazier picked his examples is clogged with weeds. They are ever present, some actively poisonous, some seriously time-wasting, some hardly worth bothering about. Pseudoscience, antiscience, the paranormal, astrology, homeopathy, Frazier enumerates and describes each species one by one, 157 of them. They range from those that absolutely cannot be true such as perpetual motion at one extreme, to graphology at the other. Graphology's claim to diagnose personality from handwriting easily could be true, is even plausible, but unfortunately is not supported by evidence. Somewhere in the middle of the spectrum we find the likes of dowsing, acupuncture, and the Loch Ness Monster. Frazier goes out of his way to disclaim debunking as the primary motive of the skeptical movement of which he was a leader and I a foot soldier. Instead, we will investigate paranormal claims—skeptically but giving those claims their best shot at success.

That movement has an interesting history, and nobody was better placed to chronicle the inside story than Kendrick Frazier. His chapter "The Rise of Organized Skepticism" takes us back to the mid-1970s when the world "was awash in unexamined paranormalism. Astrology was in high vogue (the 'Age of Aquarius'). 'What's your sign?' passed for a mainstream conversation starter." Lalla Ward told me she overheard a naïve young starlet approach Otto Preminger, distinguished director of the film they were both working on. "Oh gee, Mr. Preminger, what sign are you?" "I am a Do Not Disturrrb Sign" was his immortal reply, in a thick Austrian accent. It was the time when *Chariots of the Gods* was a major bestseller (a dear relative of mine was completely taken in). Serious scientists were fooled by a spoonbending charlatan, and it took another conjurer, James Randi, to unmask him as an ordinary conjurer masquerading as supernatural.

A loosely knit consortium of Americans saw the need to combat this wave of credulous nonsense: Carl Sagan, Martin Gardner, Isaac Asimov, Ray Hyman, James Randi, and others came together, assembled by the organizing talents of Paul Kurtz. CSICOP, the Committee for the Scientific Investigation of Claims of the Paranormal, became CSI, the Committee for Skeptical Inquiry, and its journal the *Zetetic* became the *Skeptical Inquirer* with Kendrick Frazier himself as editor. Paul Kurtz led the movement, along with the Council for Secular Humanism (CSH), America's leading humanist organization. They merged under the umbrella of CFI, the Center for Inquiry, on the board of which I came to know and admire Kendrick Frazier.

Ken was not only one of America's great editors, he was a superb writer himself. I started to read this book and finished it in a day—

something I can't often say of a book. After four decades in the editor's chair, a normal individual might pardonably have settled into a comfortably jaded rut. In Ken's case, not a bit of it. Paul Fidalgo's obituary as editor of *Free Inquiry*, CFI's other journal, got it exactly right: "Not only was Ken not bored or jaded, he was ebullient. As he assembled each new issue of *Skeptical Inquirer*, he would be bursting with pride about the wealth of knowledge and insight he had the privilege of sharing with readers. He was so grateful that his job allowed him to keep learning fascinating new things every day, all the way into his ninth decade on Earth." I couldn't say it better, so I won't. Except to add that he was one of the most humanely decent men I ever met.

INTRODUCTION
Escaping the Maelstrom

Misinformation swirls about us like a hurricane that never ends. It is constant, ever-renewing, resilient, overwhelming. It seems that we might never escape the maelstrom. Falsehoods have always flown like the wind, while truth does a slow walk, as Jonathan Swift and others have long noted. But something seems different these days. Way back in 1987, my organization held a conference called "The Age of Misinformation." That theme was both topical and prescient because today, in this third decade of the twenty-first century, studies show that misinformation spreads faster, farther, broader, and deeper than accurate information. The algorithms of social media sites, the bots of bad-actor nations, and the current political-cultural climates of divisiveness all actively encourage the spread of misinformation. These and other toxic social forces amplify the abuse exponentially and poison our own sense of reality and the trust in others necessary for societies to cohere and for democracies to function.

When that misinformation and disinformation seek to nullify facts and evidence about the science of nature, life, and ourselves and present false and unsupported views as true, the result is something we can likely call pseudoscience. Pseudoscience is everywhere, following real science like a shadow, never quite revealing itself for what it is. Pseudoscience is pervasive, potent, unrelenting. It confuses people and impedes the public acceptance of good science. It advances powerful countercurrents contrary to common sense and good science, making truth constantly swim upstream against cascading headwaters of misguided misinformation designed to appeal to our deepest fears, wants, and wishes. What can be done? The first thing is to recognize it. To begin our inquiry, let's consider some hypothetical but all-too-typical situations.

Your son-in-law's brother is smart and educated. He is an engineer, but he has a tendency to embrace strange conspiracy theories. He proudly propounds ideas that don't sound credible. When the COVID-19 pandemic hit and vaccines against this terrible ailment became widely available, he refused to get vaccinated. He somehow managed to get a religious exemption, even though he is not religious. The family quit inviting him to holiday gatherings, first because they became tired of hearing his exaggerated stories, and now because he was unvaccinated.

You like a work colleague, but he sees conspiracies everywhere. When the pandemic hit, he embraced many of the myriad conspiracy theories about how it happened, and later he fell in with the virulent antivaxxer movements that encouraged vast segments of the population to remain unvaccinated, endangering everyone else. He still thinks that the collapse of the World Trade Center buildings on 9/11 was brought about by explosives purposely hidden inside the buildings, not from the hijacked airliners that crashed into them. He even still has some doubts that the United States landed men on the moon. He and his wife worry about fluoride in our water. He sometimes talks about ancient orders of secret societies he thinks somehow control our destinies today. He goes on social media and finds welcoming groups devoted to all kinds of conspiracy theories. They reinforce his leanings, and he soon seems swallowed up by these rabbit holes of misinformation and near-paranoid thinking.

Your neighbor is educated, smart, hardworking. He's a regular guy in most respects, except for one thing: He's convinced that beings from another planet have visited Earth. Not only have they visited, but they also have interacted with the population and caused certain people to undergo personal "transformations" that are so profound that they have achieved a higher plane of reality. This belief your neighbor holds so adamantly seems incongruous for someone you like and care about. You've talked with him about it and gently questioned why he believes what he believes, but he seems terribly sincere and sure. You realize this isn't a casual fad he has momentarily adopted but a lifetime emotional commitment—an entire personal worldview shaped by what to your mind might seem a delusion.

You are a biology teacher with a very bright student. She has studied all the material in your unit on biological evolution and seems to understand it. She can pass your tests and parrot back to you the basic facts and concepts, but you can tell she has serious reservations about the subject, so much so that you wonder if she is studying it only to develop arguments that she may use later to try to oppose its key insights.

Your dear friend recently was diagnosed with cancer. Without treatment it is likely fatal, but she decided to avoid any treatments that Western medicine offers. Instead, she underwent a variety of alternative therapies—colonics, specialized exercises involving expensive equipment she had to buy, and magnetic therapies. Her healers told her that by doing so she will avoid the poisons of chemotherapy and radiation and be cured of her cancer. She accepted their advice and avoided scientific medical treatments for two years, with almost predictable results.

What can you do? Most people start looking for sources of reliable scientific information, but how do you know what's reliable? How, in fact, do you know what really is scientific? And what does that mean anyway? How can you tell good science from bad science, real science from bogus science? We know science can do enormous good, but is there a way for regular people to judge what is real science and what isn't? And if you are a scientist, physician, teacher, engineer, or someone else who already has considerable scientific knowledge, then how can you best deal with this gap between what you know about science and the troubling misconceptions held by your students, patients, clients, neighbors, and much of the general public?

I hope this book can help. This book is about thinking about science and thinking about pseudoscience. *Pseudoscience* is a term mostly unfamiliar to us. I suggest it should be better known and understood. I want everyone to be able to identify pseudoscience more easily. The differences between science and pseudoscience are immense, but in practice, most people need some help seeing them. There are good reasons the task is difficult. I examine them in the course of this book.

Why not call it *fake science*? That term is shorter. It has some power. Some things I present about are definitely fake science, but I prefer *pseudoscience* because the term *fake science*, like *fake news*, has already become politicized and brandished like a weapon by partisans of particular points of view, without regard to science, facts, or evidence. *Pseudoscience*, in contrast, has a long history of use within science and philosophy. It has not been coopted by political partisans.

First, about science: I love science. I love its exuberant quest to find out things. To answer our questions about nature, the universe, ourselves. To have the tools to out what's real and what is not. To find out what is most likely true and not true. Science is one of humanity's greatest and most successful inventions. We should exalt it as much as we do music, art, literature, sports, and entertainment. It is a treasure—an underappreciated treasure. (Yes, it has flaws, as well, like any activity carried out by human beings, but science works mightily to minimize them. I talk about that later.)

When asked, large majorities of people say they trust science. Even in this age of distrust, polls consistently show that trust, both in the United States and globally. In the United States, science is trusted more than any other group except the military; in other countries the military ranks farther down. Most people *do* appreciate science's findings, its major discoveries, and they definitely appreciate its practical uses, like developing new medicines and treatments and new cell phones that have amazing qualities. Many others, though, still stubbornly reject certain findings they feel conflict with their own worldviews. Nevertheless, few know much about how these advances were achieved, the kind of thinking and effort behind them. Science's innovative methods, its emphasis on systematic observation and experiments, its safeguards, and its communities (collaborative *and* competitive) provide a unique blend of powerful creativity and systematic rigor. Together, all these attributes allow us to winnow out deep truths about nature and distinguish reality from conjecture. In these weird and troubled times, when fact and opinion, evidence and belief, fake news and real news are increasingly conflated and where lies and fake information spread on social media far faster, farther, and deeper than truths and real news, this ability to distinguish what is real is more important than ever. It is crucial to the health of our democracies and to our long-term survival.

In thinking about these questions, we need to make a distinction between science and scientists. Scientists are human and imperfect. Like all of us, they make mistakes. They may have strong views and personal biases. As Isaac Asimov once said, "A scientist is as weak and human as any man, but the pursuit of science may ennoble him even against his will."[1]

Scientists are taught and strongly encouraged to try to overcome these personal influences and to favor the evidence. They are trained to seek objective facts and shun subjective beliefs—difficult as that may be to do. But we don't just rely on their good intentions. Hardly! The processes of science have something built into them that few other human activities have: a series of error-correcting mechanisms. These, too, are not perfect, but over time, they tend to work very well. They help root out provisional findings that don't stand up to repeated tests. They help distinguish solid scientific results from wobbly ones. They sift out results that are most likely inaccurate and invalid from those that are most likely true and real. They are our *reality detectors*. They are our bullshit detectors. And over time, this leads to an increase in real knowledge. That is the special and mostly unappreciated genius of science.

These processes are messy. There are ups and downs. There are wrong turns and dead ends. They take time. Scientists are generally aware of how they work (or sometimes don't), but to the rest of us, they remain essentially invisible, lost in the fine details that seldom make the news. And while scientists are indeed imperfect, most good scientists do their best to take a careful and even humble approach to nature's almost unfathomable mysteries. This contrasts sharply with those who claim they seek knowledge but don't embrace science's methods and values. They instead tout amazing claims before anyone else has confirmed them and sometimes even after many others have *dis*confirmed them. When they do so, they may be fooling either themselves or the rest of us. This counterscience is ubiquitous. It has a beguiling appeal. We call it pseudoscience.

Pseudoscience is everywhere, lurking in the shadows of real science. Pseudoscience pretends to be science and coopts some of science's language, but it betrays itself with an absence of scientific methodology, grandiose and unsupported claims, direct appeals to the public, reliance on anecdotes and testimonials, rewards for wishful thinking, and failure to build on recent and past published scientific research. Antiscience, a special case of pseudoscience, is a direct hostility toward science or toward unwelcome scientific findings. It often has organized support. Antiscience has several especially pernicious strains in today's public discourse: climate science denial, the antivaccine movements that dangerously slowed our battle against the COVID-19 pandemic (and other diseases before that), antievolution creationism, opposition to GMOs—the list is long.

Pseudoscience and its antiscience components confuse the public and impede scientific progress. Unchecked, they are dangerous to ourselves and to our democracies, which (I hope you agree) depend on our citizens and political leaders knowing the difference between what's real and what isn't. Facing up to pseudoscience and antiscience poses important scientific, intellectual, and practical challenges that are far from straightforward. Exploring the blurry boundaries between science and pseudoscience continues to fascinate philosophers, scientists, and vast segments of the general public alike.

As this book shows, all scientific fields have pseudosciences living in their shadows, mimicking them in various ways and seeking the reflected respect science has earned. From time to time, the quieter ones get their opportunity to emerge and to try to make you think they are just as real and valid. Others preen with flash and glitz, have their own TV shows and websites, and promote themselves and products with brashness and certitude. This is the case whether you are talking about physics and astronomy, the earth sciences, anthropology and archaeology, psychology and the social sciences, or biology and biomedicine (the latter now especially).

In these times, medical pseudoscience, or *pseudomedicine*, has emerged to dubious supremacy. It is a big part of what proponents call alternative medicine. Those who know better realize that pseudomedicine—pseudoscience in health care—is perhaps the largest single field of pseudoscience today, and the one most filled with potential harm to consumers. Even more than many other pseudosciences, pseudomedicine has a vast reach and influence on our media, politics, and culture. Pseudomedicine has attracted famous celebrity promoters. Some are physicians themselves with a weird sense of their responsibilities under the Hippocratic oath; others are television or movie stars who unflinchingly tout dubious products for profit. But pseudomedicine disguises itself even better than most other pseudosciences and so has gained a semirespectability it does not deserve.

I've professionally observed, studied, and confronted pseudoscience for a very long time. As a young science journalist, I was editor of *Science*

News magazine in Washington, DC, where we tried to bring accurate and clearly explained scientific information to the public weekly. We did our best to present new advances in a scientifically responsible way, in perspective and without exaggeration or sensationalism. Before that, I was editor of the newsletter of the National Academy of Sciences, reporting on panels of the country's greatest scientists called to provide advice on pressing national issues; even there, we occasionally dealt with fringe claims, such as publishing the academy's formal review before its official release of the famous Condon report that cast a negative light on claims of flying saucers. And for the past four decades, I have been editor of the groundbreaking bimonthly *Skeptical Inquirer: The Magazine for Science and Reason*. Many of the world's most prominent scientists, scholars, educators, and investigators have contributed. They examine all manner of extraordinary claims, usually those having some scientific angle or content, even if not using a scientific approach. The involvement of these scientists and academics brings high standards of scholarship and raises the level of discourse. They bring scientific rigor and thinking to an area sorely deficient in those qualities. Their efforts reemphasize to us all how interesting, significant, and relevant these examinations are. Like them, I find these subjects both inherently fascinating and crucial.

In this book, I want to give a sense of science's methods, aims, and values and forthrightly contrast them with those of pseudoscience. I report on some of science's recent discoveries and show a bit about how they were achieved. I can't do the same for pseudoscience because—well—it has none. You can understand pseudoscience only if you understand something about real science. I examine the myriad issues and competing ideas encountered in identifying and confronting pseudoscience (including whether we should even use that term; I argue that we should). I consider the harm it does (to people, to society, to science itself, even to the animal world); chronicle scientific efforts to counter or expose it; and survey some prominent examples of investigations, past and present. I offer a full chapter on the values of science, a topic I think is too little appreciated. That is probably because most scientists are loath to talk to the public about their values because doing so may seem self-righteous or self-congratulatory. I can do that for them. Because I have covered weather and climate science all my life, my extended case study on antiscience focuses on climate antiscience and denial. I also provide one of the first extensive inside reports on the rise of the modern skeptical movement that examines pseudoscience and extraordinary claims of all sorts, first in the United States and then globally. I provide numerous examples that demonstrate how pseudoscience, pseudoscientific thinking, and antiscience infect the wider culture—all of human society, in fact, including, most recently, our political systems. I write in a personal style that draws on my own experiences with these subjects. I hope you will find it congenial. I do my best to avoid jargon and undue complexities.

In the end, I hope professionals in these fields may find some new ways to look at these troublesome issues. I hope to reinforce their commitment to critically examining pseudoscience. I hope to provide some needed structure and overview to what can at times seem a fragmented, isolated activity. And I hope to raise some new issues for discussion. For the intelligent general reader who shares my sense that science is wonderful but who nevertheless is curious about fringe science and its extraordinary, all-too-wonderful claims—or who knows a loved one who has fallen prey to the siren songs of pseudoscience—I hope that you may see why I love and treasure science so much. I hope you will be better able to identify the pseudoscience and antiscience swirling all around us and see how they have managed to infiltrate our institutions so pervasively. My hope is that this may equip you to come up with your own ideas and strategies to help others in this quest.

And when you are all done, I hope you all may share some of my passion, love, and appreciation for the methods, values, ideals—and ultimate effectiveness—of science. If you weren't already, I hope you may become newfound or newly committed defenders of science and reason.

CHAPTER ONE
SCIENCE AND THE FRONTIERS OF DISCOVERY

Many people have an unfortunate misconception about science. We might call it the textbook view. It holds that science is dry, static, dull, settled. Scientists don't think that. I don't either. It is an understandable but misguided view. It probably stems from our school days. I can sympathize with it and see where it arises. I am sure you can, too. When you learn some of what's happening at the leading edge of scientific discovery, though, all that can change. You suddenly enter a fascinating world of novelty and delight. New findings in science can invigorate us. They can inspire us. They can give us a sense of progress and possibility. They lead to new ideas and innovation. They change the world.

This exciting new world is *initially* more tentative, less certain, to be sure. It requires discernment. Hundreds of thousands of scientific papers are published each year. Many prove incomplete or insufficient in some way and more or less fall by the wayside. Most advance their science in small, incremental, yet necessary steps of importance, mainly to their fellow researchers in a particular sub-subfield. A few take bigger leaps and have the prospect of contributing mightily to our larger scientific understanding of the world. Yet always—yes, always!—the findings are subject to revision, from new and better evidence, more advanced instrumentation, challenging critiques, and different and more insightful interpretations. That openness to new evidence is a powerful strength of science.

This openness is the opposite of dogma. Dogma, resistance to all contravening evidence in favor of a deep-seated belief, is anathema to science.

To start this book and I hope to get us all thinking, here are a few recent scientific results I find especially interesting. Where along this spectrum of significance I mentioned earlier do you think these might fall? And where might they end up? In future textbooks? Forgotten? Or somewhere in between, incremental contributions to scientific understanding?

Are There Earthlike Planets Nearby?

Thousands of planets orbiting other stars have been tentatively identified. We call them exoplanets. Those discovered first were naturally very large and probably Jupiter-like gas giants. But in 2016, scientists using the European Southern Observatory's planet-searching telescope HARPS announced detection of a rocky planet about 1.3 times the mass of Earth orbiting Proxima Centauri, a red dwarf star just 4.24 light-years away in the triple-star Alpha Centauri system—the nearest to Earth. The star is a lot different from our Sun, however; it has a magnetic field six hundred times stronger than the Sun and periodically erupts in massive flares, which might make any possibility of life developing on the planet remote.

Then, in 2017, NASA and a Belgian-led research team announced discovery of not one but *seven* Earth-size rocky planets around *one star*, Trappist-1, thirty-nine light-years away. This star is a cool and dim dwarf star, but the newly discovered planets orbit it close enough to get suitable light and heat. Three of the planets, in fact, are in the habitable ("Goldilocks") zone, where liquid water, and therefore life, might exist. If this kind of configuration is common, one astronomer commented, then our galaxy could be teeming with Earthlike planets. This announcement was greeted with enormous enthusiasm. Some called it hitting the exoplanet jackpot.

The Discovery of Gravitational Waves

Einstein's general theory of relativity predicts the existence of gravitational waves, ripples in the very fabric of space-time created by colossal cataclysms in the universe. For four decades, scientists looked for them in vain. Many alleged discoveries proved to be false alarms. Then, in the early morning hours of September 14, 2015, something incredible happened. LIGO, a sophisticated gravitational wave detector in Louisiana, had just been upgraded so that it could discern a change of less than one one-thousandth the diameter of a proton in the four-kilometer-long distance between its mirrors, and it was turned on for the first time. Just an hour later, it got a signal. Seven milliseconds later, an identical LIGO detector in Hanford, Washington, three thousand kilometers away, got exactly the same signal. The waveform, or pattern of vibrations, recorded at the two sites (indicating how much the two detectors' tunnels lengthened and shrunk as the wave passed) was identical. This was the long-sought, first direct detection of gravitational waves.

Analysis of the waveform showed that this disruption in space-time was caused by the collision of two massive black holes 1.3 billion light-years away, one thirty-six times the mass of the Sun, the other twenty-nine times the mass of the Sun. In two-tenths of a second, the two black holes merged into one single black hole sixty-two times the mass of the sun. The remaining three solar masses were converted into gravitational wave energy, so much so that it shook space-time and was detectable here on Earth.

For five months, the LIGO scientists carefully scrutinized the data to make sure they were right. They held their breath while reviewers of their draft scientific paper checked everything. Finally, in late January 2016, the reviewers' feedback came in: They were positive, enthusiastic, even congratulatory. When the discovery was announced in Washington, DC, and simultaneously published in *Physical Review Letters* on February 11, 2016 ("Ladies and gentlemen, we did it!" exclaimed LIGO's executive director), the world scientific community was jubilant. The news media and public joined in the celebration. The editors of *Science* magazine chose the discovery of gravitational waves as its 2016 breakthrough of the year. Gravitational waves from three more black hole collisions have been detected since. As I expected, when the 2017 Nobel Prizes were announced, the prize for physics went to three of the leaders of the LIGO project for this discovery.

By 2021, five years after the discovery of the first gravitational waves, more than fifty additional gravitational-wave events had been detected, dozens of them from binary black hole mergers (others are from collisions of neutron stars or other compact objects). The two original gravitation-wave detectors underwent upgrades to their sensitivities, and additional detectors were under construction, so we can expect still more exciting discoveries in the future.

The First Image of a Black Hole

In April 2019, an international team called the Event Horizon Telescope (EHT) Collaboration announced they had obtained the first image of a black hole. To make this epic accomplishment, they used ten radio telescopes on four continents to effectively create a single, virtual telescope the width of the Earth (which provides the extremely small resolution needed). The achievement confirms (once again) the predictions of Einstein's general relativity theory.

The project required immense planning and cooperation and was only possible because they chose to image a *supermassive* black hole, the one long suspected to be in the center of a galaxy called M87. Even though M87 is part of our "local universe," it is still 55 million light-years away. But it turns out that the black hole at its center is so massive that it holds the equivalent of 6.5 billion of our Suns. Its diameter of 38 billion kilometers (24 billion miles) is almost the size of our solar system. The ring of light we see in the image is electromagnetic radiation emitted as a disk of accreted matter surrounding the black hole as it falls into it. The black hole itself is indeed black.

The EHT Collaboration has also been attempting to image the much more ordinary-sized black hole that was assumed to be at the center of our own galaxy. It is in a dimmer source called Sagittarius A* and is only one one-thousandth the size of the M87 black hole. Counterintuitively, it is much more difficult to image than the far more distant M87, not just because it is smaller, but also because of the dust and gas on the galactic disk. Finally, on May 12, 2022, came the announcement: The EHT team had captured the first image of the black hole at the center of our own galaxy. The long-awaited image looked very similar to the image of M87. It was the first direct evidence that the object at the center of our own galaxy is also a black hole. It has about 4.15 million times the mass of the Sun. "What's more cool than seeing the black hole at the center of the Milky Way?" said one of the former EHT team members at a Washington, DC, press conference.

How Old Is the Universe?

The European Space Agency's Max Planck space telescope recently measured the universe's primordial cosmic background radiation to enough precision to determine that the universe is 80 million years older than had been thought: 13.8 billion years (rounded to the nearest significant figure) instead of 13.7 billion years. More specifically, the universe is 13.798 billion years old, give or take 0.037 billion years! The same observations also show that the universe is 3 percent broader than we'd thought and is expanding at a rate 3 percent less than previous estimates. To accomplish these findings, the Planck satellite had to measure temperature variations in space caused by the afterglow of the big bang to one one-millionth of a degree. One physicist said of the results, "Amazing."

What Is the Universe Made Of?

The same Planck satellite measurements show that the universe is made of slightly more matter and slightly less energy than previously thought. In the lifetimes of most of us, we have learned that the ordinary matter that makes up everything we are familiar with can account for only a small fraction of the mass and energy in the universe. *Dark matter* is an unseen *something* (we know it only from its gravitational effects) that holds galaxies together but doesn't interact with light. Then, in 1998, scientists discovered that the expansion of the universe is happening at an ever-increasing rate. It was an epochal finding. It brought Nobel Prizes to its discoverers. *Dark energy* is the name we've given whatever it is that is making that happen.

Before the Planck observations, we thought the universe's makeup was 72.8 percent dark energy, 22.7 percent dark matter, and 4.5 percent ordinary matter. The after-Planck totals are 68.3 percent dark energy, 26.8 percent dark matter, and 4.9 percent ordinary matter. Those may not seem very big differences, to be sure, but consider that not too far back, we had no awareness of dark matter and dark energy at all. Think what it took to gain the ability to determine these proportions, let alone to 0.1 percent. And then think what it means to say that almost 95 percent of the mass-energy in the universe is still unidentified. Some big discoveries must lie ahead. How exciting! How humbling!

Has a Signal of Dark Matter Been Detected?

Dark matter can't be seen, of course, but two groups of physicists in 2015 reported detection of an X-ray signal that they suggest might be due to the decay or mutual annihilation of particles of dark matter. They reported X-ray emissions having an energy of 3.5 kiloelectron volts (keV) coming from several galaxies, including our own, and galaxy clusters. The journal *Physical Review Letters* accepted a paper by one group from Leiden University in the Netherlands reporting such a peak coming from the center of our own Milky Way galaxy. The authors said the intensity of the 3.5 keV peak lies in the right range to be produced by dark matter reactions. But other researchers were skeptical that the emission lines were real or that dark matter collisions could be the only explanation for them. Better observations by new X-ray satellites were needed to plot the shape of the peak and strengthen the case for these emissions being flashes of dark matter.

What Caused the Biggest Mass Extinction in Earth's History?

About 252 million years ago, some catastrophic event happened that destroyed 90 percent of all the marine species on the planet, leaving us a fossil record of near-total destruction that geologists call the Permian-Triassic boundary. The cause of this close-to-total pinch-off of life on our planet has long been a mystery. In late 2013, Earth scientists reported that they had identified the likely culprit, and it confirms previous speculations. A series of massive volcanic eruptions in Siberia lasting two million years spewed volcanic sulfur dioxide into the atmosphere, touched off massive coal fires, carpeted an area of Siberia the size of western Europe with basaltic rock (geologists call that area the Siberian Traps), and nearly brought life on Earth to a close. Geochronologists recently dated the events to a precision never before possible using the slow but steady decay of uranium-238 to lead-206 in crystals. They found that the eruptions began 252.28 million years ago, and the extinctions began 251.94 million years ago, properly placing the putative cause ahead of the undoubted catastrophic effect.

It had to be a momentous transition. The first organisms on Earth were single cells. Getting from that stage to more complex multicellular organisms—which has enormous advantages—has always seemed a giant hurdle, an almost unbridgeable gap. But recent findings indicate the gap may not have been so unbridgeable after all. First, studies of evolutionary history show that the switch happened multiple times in the past, and this indicates the hurdles must not have been so high as thought. Second, genetic comparisons now show that much of the genetic equipment to change from single-cell to multicell status was already in single cells' genetic package before the leap started. And third, new laboratory experiments have shown that the transition to multicellularity can happen in just a few hundred generations, nothing at all in evolutionary time. So what seemed a difficult biological hurdle now seems not so much of an unexplainable leap after all.

What Did We Learn from Landing on the Bed of an Ancient Lake on Mars?

NASA's Mars Perseverance rover landed on the floor of Jezero crater on Mars on February 18, 2021. The site was chosen especially because orbital images had shown two sedimentary fan structures at the site inferred to be river delta deposits from an ancient lake on Mars (3.6 to 3.8 billion years ago). Two cameras on the rover took a series of detailed images of the outcrop faces of the western fan, invisible from orbit, during the first three months after landing.

In November 2021, an international team of thirty-nine scientists reported the results. They show that the lake, which they determined had no outlet, underwent two different hydrologic periods. The first was fairly normal, with inflows of water forming river deltas much as they do on Earth. But the second, indicated by conglomerates of boulders in the uppermost strata, was more violent and episodic, with floods so energetic they were capable of moving meter-size boulders over distances of potentially tens of kilometers. These multiple flood episodes could have been due to intense rainfall, rapid snowmelt, or even built-up glaciers surging episodically. And the finer-grained, bottommost strata, they say, has "high potential to preserve organic matter or potential biosignatures."[1] So humans, via their instrumented spacecraft surrogates, have landed on an ancient Martian lakebed and helped decipher its evolution, confirming that Mars was once warm and humid enough to support a hydrologic cycle, at least episodically. As for whether these conditions could have supported life, stay tuned.

Where Is the Second-Tallest Mountain in the Solar System? What Dwarf Planet Has an Ocean?

We know that Olympus Mons, on Mars (subject of an Arthur C. Clarke short story, "The Snows of Mount Olympus"), is the largest mountain in the solar system, rising twenty-two kilometers, or nearly fourteen miles, above the Martian surface. But what planet has the second-highest mountain? Answer: Not a planet at all (at least not yet). It's an asteroid, Vesta. NASA's Dawn spacecraft orbited Vesta in 2011–2012 and surveyed the surface. It found a mountain in a large impact basin at the asteroid's south pole towering twenty kilometers, or more than twelve miles, above its base. This makes the mountain larger than Mauna Kea, the largest mountain so measured on Earth, which rises ten kilometers above the seafloor. The Dawn satellite also showed that Vesta has ancient lava flows and tectonic features. As a result, Dawn's principal scientist said he now considers Vesta (530 kilometers in diameter) to be the smallest terrestrial planet, putting it into the same category as Earth, Mars, Venus, and Mercury.

Dawn then flew on to dwarf planet Ceres, which comprises a third of the total mass of the asteroid belt, and went into orbit around it in July 2015, thereby becoming the first spacecraft to orbit two planetary bodies. From then until November 2018, Dawn studied Ceres from orbit, coming as close as twenty-two miles from its surface and making more epic discoveries. One of them, announced in August 2020 in the journal *Nature Astronomy*, was especially amazing: evidence of a briny ocean. Bright spots seen on Ceres from early photos subsequently turned out to be a crust of sodium chloride—salt—deposits, and later, high-resolution images from the close orbit led to the discovery of an underground reservoir perhaps hundreds of miles across lurking twenty-five miles below the dwarf planet's Occator Crater. "Ceres is now an ocean world," declared one planetary scientist. The salt deposits and the water bringing them to the surface are young, so it looks as if Ceres not only has one or more subsurface oceans but also that the planet may still be quite active. It could even be a good place for the synthesis and catalysis of complex prebiotic chemicals.

Can We Make Cancers Destroy Themselves?

Typically, discoveries are not sudden but accumulate over time from the work of many research groups. That is what has happened with cancer immunotherapy. There is a molecule called CTLA-4 that keeps our cancer-fighting T cells from doing their job. Evidence has been building for some time now that it may be possible to block the action of CTLA-4—in effect blocking the blocker—and thus springing the immune system free to destroy tumors. Clinical trials are beginning to show the success of this approach. These and other positive signs have grown to the point that a leading science journal chose cancer immunotherapy as 2013's breakthrough of the year.

What Do Our Molecules Look Like?

We now finally can obtain images of important large molecules in our cells. In 2014, scientists at Cambridge University reported that they have obtained images at near-atomic resolution of large macromolecules. By averaging thousands of electron cryomicroscopy images before the process damages the cells, they have imaged the structure of a large subunit of the mitochondrial ribosome. Ribosomes translate the linear genetic code into three-dimensional proteins. The resolution is an incredible 3.2 angstroms; one angstrom (10^{-10} meter, or one-tenth of a nanometer) is approximately the size of a hydrogen atom. The achievement is said to herald a new era in molecular biology. Even a few years earlier, what these scientists achieved was thought near-impossible.

Were Humans in Europe 200,000 Years Ago?

In July 2019, a team of twelve researchers reported that two skulls from Apidima Cave in southeastern Greece date to more than 170,000 years ago and indicate that two different forms of human groups occupied the site. Their uranium-series radiometric dating methods and their analyses of the damaged skulls determined that one skull is more than 170,000 years old and has Neanderthal-like characteristics. But the other skull is even more surprising. It dated to more than 210,000 years ago, and they say it presents a mixture of modern human and primitive features. The researchers contend that this indicates that both an early *Homo sapiens* population and a *later* Neanderthal population were present in the cave. If this finding holds up, then the 210,000-year date would be the earliest example of *Homo sapiens* outside the African continent. The date also precedes, by 160,000 years, any *Homo sapiens* fossil previously found in Europe.

The researchers said their findings support multiple dispersions of early modern humans out of Africa and show the highly complex dynamics between different populations of modern humans in southeast Europe. The skulls have been in a museum in Greece since 1970, and a number of paleoanthropologists not involved in the research urged caution toward the claims.

I find it exhilarating that we can learn such things. We are primates. Our ancestral cells evolved in Earth's primordial muck, and our upright-walking ancestors emerged out of the African savannahs only a few million years ago. I find it moving that since that time, we have not only developed the curiosity about the origin, the age, the size, the physical makeup, and the future of the universe (and ourselves) but also have very recently created the intellectual and technological tools to find the answers—and to comprehend those answers once found.

Multiply these discoveries and insights by many thousands in every scientific field and then by the ever-increasing number of scientific fields and subfields, and you may begin to get a sense of the vigorous, robust world of scientific discovery before our very eyes. Yet most of us have only the vaguest awareness of it, and that's even after you subtract the results that will be found flawed or wrong in some way, a natural and expected part of science at the frontiers.

Regarding flawed results, let's take one more example of a fairly recent discovery: *Have we detected the earliest evidence of cosmic inflation?* On March 17, 2014, American scientists using a special telescope in Antarctica announced, to enormous worldwide excitement, that they had detected the earliest echoes of the big bang. They announced that their BICEP2 telescope experiment had detected gravitational ripples in the cosmic background radiation generated from the amazingly short but incredibly rapid period of exponential cosmic inflation that followed the big bang explosion that created our universe.

The discovery was widely heralded. Some scientists compared it in importance to the 2012 discovery at the CERN Large Hadron Collider of a long-sought particle called the Higgs boson, which explains the origins of mass. That discovery garnered the 2013 Nobel Prize in Physics for the theoreticians who predicted it (François Englert and Peter Higgs). Yet only two months after the exciting news of the cosmic inflation discovery, in May 2014, came disappointing news from a different research group: What the Antarctic telescope had detected may have been an artifact of dust within our own galaxy. The size of the signal in the cosmic background radiation had been a surprise, but it now looked as if a misunderstanding about complex maps of the intervening dust may have led scientists to overestimate and misinterpret the signal.

Their research wasn't yet published in a scientific journal, and they were awaiting new data from the previously mentioned Planck satellite. That came September 22, 2014, and it cast further doubt on the initial claim. It was a detailed map of interstellar dust in our galaxy. Planck's full-sky map shows that the patch of sky observed by the BICEP2 telescope was not among those areas with the least dust. In fact, the new results showed there are no "clean" windows where primordial cosmic microwave background radiation can be measured without subtracting emissions from dust. And this means it is even more likely that the initial result was due to such dust and not a signature of primordial gravitational waves. A joint analysis by both groups began and was expected finally to resolve the dust-contamination issue one way or the other.

In early 2015, the results of that collaborative analysis were announced. It was a near death knell for the original discovery. The BICEP2/Keck Array and Planck Data collaboration group consisting of more than two hundred scientists found that the effects of the intervening dust were such that the significance of the original data is too low to be interpreted as evidence of primordial gravity waves. This means that the original data from the Antarctic telescope were almost certainly not a detection of gravity waves from the big bang. "Unfortunately, we have not been able to confirm that the signal is an imprint of cosmic inflation," conceded Jean-Loup Puget of the Institut d'Astrophysique Spatiale in Orsay, France, and principal investigator of the Planck instrument.[2]

If wrong, as appears to be the case, then is this finding an example of bad science, flawed science, or just regular science? Well, the scientists in question and their institutions all have good reputations. They were using extremely complicated equipment and attempting detection of something never before detected. They certainly thought they were being very careful. It is hard to fault them too much. In an important sense, this episode, still in progress, shows how the processes of science work, bringing intense critical scrutiny to every new finding. Only those findings that successfully pass through all these critical filters repeatedly can become part of human knowledge. In that sense, what happened may show once again the strength of the scientific process in ensuring that any new claim to knowledge has repeatedly survived rigorous examination.

Here's another example, even more recent, of the filters in action. They are not pretty in this raw form, but the result should eventually bring greater clarity. *Did a geomagnetic event 42,000 years ago change life on Earth?* An international group of thirty-three scientists published a paper in February 2021 based on studies of a 42,000-year-old kauri tree trunk recently unearthed from a New Zealand bog where it had been nicely preserved. Their studies showed that radiocarbon levels in the wood's tree rings surged around that time, an indication that Earth's protective magnetic field weakened and its magnetic poles flipped, north to south. Such reversals have happened many times in Earth's past and have been thoroughly documented, but this one is geologically very recent, during human prehistory.

They pinned down this magnetic flip event, the most recent in Earth's history, in fairly fine detail. But in their paper, "A Global Environmental Crisis 42,000 Years Ago," the scientists, led by Alan Cooper of the South Australian Museum, went well beyond that. They speculated that the increases in incoming radiation (from cosmic ray bombardment) would have briefly shifted Earth's climate, contributing to the disappearance of large mammals in Australia and the Neanderthals in Europe. They even speculated that the magnetic event may have had something to do with humans' beginning to make elaborate cave drawings in Europe and Asia. If true, then this discovery would be a major advance in understanding human prehistory. Even at the time of publication, some scientists expressed concerns about the group's speculations.

Then nine months later, in November, two different teams of scientists blasted that part of the paper. Their full texts were published online in the form of "technical comments," but brief abstracts of them were published in in the same journal as the original paper. John Hawks's critical comment ended, "These authors misrepresent both the data and interpretations of cited work on extinctions and human cultural changes, so the specific claims they make about extinctions and cultural changes are false."[3] The second team of scientists reviewed the original group's claim about major behavioral changes within prehistoric groups and human and animal extinctions and had this biting comment: "Other scientific studies indicate that this proposition is unproven from the current archaeological, paleoanthropolgical, and genetic records."[4] Whew! One wouldn't want to be on the other end of those criticisms.

But the original study's authors, in response to these two critical comments, defended their study. To the first they wrote, "Although we welcome the opportunity to discuss our new ideas, Hawks' assertions of misrepresentation are especially disappointing given his limited examination of the material."[5] To the second critique, they responded,

> Our study on the exact timing and the potential climatic, environmental, and evolutionary consequences of the [magnetic reversal] has generated the hypothesis that geomagnetism represents an unrecognized driver in environmental and evolutionary change. It is important for this hypothesis to be tested with new data, and encouragingly, none of the studies presented by Picin *et al.* undermine our model.[6]

When I first read the original paper by Cooper and colleagues, I found it so intriguing that I underlined key passages, clipped it out of my copy of *Science*, and saved it in a file folder (I'm old school). Identifying a geomagnetic reversal in relatively recent times is important scientifically, but if the reversal had numerous effects on life on the planet at the time, then that would seemingly boost the study to still another level of significance. But then came the two critical technical comments. The critics minced no words. What to think?

First, this is how science works. New findings are subject to rigorous criticism. They prevail only if they can stand up to them. Second, note that the critics aren't condemning the paper's main finding: confirming in precise detail a reversal of Earth's magnetic field about 42,000 years ago that lasted less than 1,000 years. What upset the paper's critics were the authors' suggestions that contemporaneous climatic and environmental impacts occurred 42,000 years ago, coincident with Earth's weakened magnetic field and immediately preceding the reversed state of polarity.

The paper said the event "appears to represent a major climatic, environmental, and archaeological boundary that has previously gone largely unrecognized."[7] The critics disagreed with those sweeping statements. But how familiar were they with all that new data? This controversy is so new that the full resolution is still in the future. Who will turn out to be right? At this point we don't yet know.

Science at the frontiers is like that. We really don't know for sure until all the facts are in, all the criticisms are weighed, other scientists consider the original paper and its criticisms, and still others subject them to new analysis—perhaps even do new studies of their own. This may be hard on the original researchers, but it is a fine pathway to the truth. And to outside observers like you and me, it makes the quest even more exciting.

Polls show that most people *do* find scientific research important for many obvious and practical reasons. Scientific research forges economic progress, drives technological innovation, creates extraordinary new medical advances, and contributes mightily to the national defense. Applied research and development advance technologies, but the initial discoveries leading to many such technological advances usually arise out of fundamental research, sometimes done for entirely different reasons and decades earlier. The quest for fundamental knowledge can lead to amazing practical consequences and in unpredictable ways.

Yet even if that weren't the case, we would—as best we can—vigorously pursue scientific discovery. It is a part of our nature. It is a big part of how we create the future. We are discoverers. The quest for fundamental knowledge is noble. It is one of the best, most telling, most honorable characteristics of our species, an essential component of our very humanity. It is, at least in a figurative sense, in our DNA, our genes. Without this desire to explore, our ancestors wouldn't have made it over the next series of hills, yet they populated the entire Earth. Without this thirst for knowledge and our ever-expanding abilities to painstakingly obtain it, we would not be who we are. The wonderful, exploratory, explanatory, error-correcting methods of science—including a vast array of tools to help sift the possibly true from the undoubtedly untrue—make possible that noble quest.

CHAPTER TWO
PSEUDOSCIENCE AND UNFOUNDED IDEAS

For almost every area of science where real science is forging ahead, there is an equivalent *pseudoscience* creeping along in its shadows, seeking to gain some reflected glory from the science it tries to emulate. Pseudosciences pretend to be science, but they are not. Pseudosciences offer bizarre ideas, sensational theories, and dubious products unsupported by any findings from real science. They can be temptingly fascinating. Some are almost charming and may be fairly innocuous. Others can be deeply dangerous and pernicious.

There is an entire world of pseudoscience out there, beckoning us in myriad tempting ways. Their claims and ideas can be very colorful and appealing. One reason for that is they are not limited to the evidence of nature. They offer sensational interpretations that go far beyond any reasonable assessment of the facts. Or they make up, or at least select, their own facts.

Here we have an alternative universe. What we have so painstakingly learned about our world through scientific discovery is casually cast aside. We instead have a world with entirely different natural laws—or, better yet, none at all. This make-believe situation allows our preconceived ideas, our desires, and our imaginations to rule. No wonder the pseudosciences have such appeal. You can pick and choose what to believe is real. It is a fantasy land.

In this alternative world (or worlds), the universe and the Earth are only thousands of years old, not billions. Life doesn't evolve through natural selection (at least not "higher" life forms); humans arrived fully formed as they are now. Unidentified lights in the sky are alien spaceships. Extraterrestrial spacecraft no science has yet detected routinely abduct people. Crashed saucers and ET bodies are stored in secret government facilities. Stone monuments built by clever indigenous peoples are attributed to ancient astronauts. The planets' positions directly influence people's lives and well-being. So-called psychics see the future or have special abilities to help police find missing bodies and solve crimes. Mysterious forces other than the four known to science guide our lives and "connect" everything in the universe. Conspiracies swirl everywhere about us (beyond the real conspiracies we know about) and are the cause of multiple tragedies and human shortcomings. Humans can call on "energies" unknown to science to heal ailments and cause uncounted marvels. Untested concoctions cure cancer and heart disease and diabetes and just about everything else. Diet pills and miracle diets effortlessly slim and shape you. Homeopathic "remedies" diluted until they contain no molecules of the hypothetically active ingredient (e.g., duck liver) and have never even been tested have gained equal status with fully researched and tested prescription medications approved by official drug agencies. Metallic wristbands boost athletic abilities. Simple dowsing devices with no working parts detect hidden explosives and the presence of bad people. Remote viewers find enemy spies halfway around the world. Huge bipedal beasts roam our woods. Unseen serpents swim our lakes and seas. New forms of energy are being suppressed by the powers that be. Perpetual motion machines violate with impunity that pesky second law of thermodynamics. And—the most understandably alluring and universally accepted set of ideas—when we die, our lives continue in some other form; there is an afterlife. Ghosts, spirits, and phantasms are not metaphors but tangible realities, and with mediums' help, we can communicate with our deceased ancestors.

Some of these ideas are merely strong human belief systems or human yearnings; they make no pretentions to be science. But every one of these ideas has ardent proponents who vigorously contend that *science* supports them and earnest devotees and followers who believe fervently that that assertion is indeed true.

Pseudoscience and pseudoscientific practices pretend to be science. They are often good enough at it to fool large numbers of people into thinking that they are. At first glance, pseudoscience may look identical to science. It adopts some of the language and other trappings of science. But usually, with just a little inquiry, we can force pseudoscience to reveal its hand. With rare exception, it is not carried out in the open-minded, communal, yet skeptically oriented spirit of science. It is the antithesis of science, deceptive (even self-deceptive) in intention, methods, or result.

Science takes us into the future and creates genuine progress in human knowledge. The pseudosciences remain substantively static and take us nowhere. Or they take us backward. They frequently change their guises, but these altered appearances lead to no more reliable results. They exploit wishful thinking and our deepest human desires. They mistake the illusory for the real. That is not only stifling to progress in knowledge, but it is also deceptive and even dangerous. It creates misinformation, misconceptions, misrepresentations, and misunderstandings. And that generates confusion among a public that—for a variety of fairly understandable reasons—is only vaguely familiar with the attributes of genuine science to begin with and therefore vulnerable to the siren sounds of pseudoscience.

Each pseudoscience has in some way its own underground subculture. These subcultures often have charismatic leaders, self-congratulatory publications and products, websites and television programs, enthusiastic promoters, and intensely devoted fans and followers. Some curry the support of scientifically naïve political leaders, which gains them entrance to the corridors of power—and to taxpayers' (meaning our) money.

And then there is antiscience. Antiscience is similar in many ways but differs from pseudoscience in one fundamental sense: Rather than mimicking science and trying to gain credibility by association, antiscience actively opposes good science. It attempts to undercut well-established scientific principles and to undermine new and important scientific findings that it opposes for a variety of ideological reasons. Antiscience can be far more pernicious than pseudoscience. Pseudoscience can sometimes be, and often is, successfully ignored. Antiscience is often funded by well-organized and deeply committed opposition groups and therefore is very effective. It can be vicious and nasty, intentionally distorting and misrepresenting the science and personally attacking scientists. It must be actively fought if good science and good decision making are to prevail in the public arena.

At least three primary fields of antiscience are prevalent today. One—antievolutionism, or creationism—has been around for decades, has morphed multiple times, and opposes the science of evolution for perceived religious and cultural reasons. Another, anticlimate science, is much newer and opposes for political, economic, and cultural reasons the increasingly clear findings that the Earth is warming and our climate is changing. Antivaxxers oppose vaccines; sow doubt and discord about this greatest public health advance; and keep their children and others from getting them, greatly endangering public health. During the COVID-19 pandemic, antivaxxers became even more active, intense, and virulent. (Another similarly active and dangerous antiscience in recent decades, especially in Africa, is the anti-AIDS-virus movement, the idea that AIDS is not caused by a virus.) Some may have been sincerely motivated by misinformation-generated concerns about vaccinations; others opposed because they hold strong antigovernment views, which also made them suspicious of national and local government mask mandates. All this counterscience contributed in major ways to the prolonged spread of the disease and to increases in hospitalizations and deaths.

All these antisciences are ideologically motivated but from different positions. Nevertheless, they have much in common, although their proponents try to deny that. It is probably not a coincidence that antievolutionism, anticlimate science, and the antivaccine movement are especially strong in the United States today, compared with many other developed countries. The opposition to genetically modified foods (GMOs), while sometimes well intended, is fueled by many antiscience ideas. So, too, is the longtime opposition to nuclear power.

Antiscience proponents propose no new science of their own. They instead hunt for what they perceive to be flaws in the science they oppose—then distort and exaggerate these usually imagined flaws out of all proportion to try to convince large audiences and constituencies that the overall science cannot be trusted. They have been especially effective in advancing that agenda in recent years. It is because I love science—real science—so much that I have devoted so much of my professional life to exploring these other worlds: pseudoscience that mimics science and antiscience that opposes inconvenient findings. Both cause immense confusion and misunderstanding.

I love science. I love it for what it can reveal about the natural world, including ourselves. I passionately want people to share some of my delight in real science. I want people to experience some of the awe I feel when I look through my telescope and track the movement of Jupiter's moons, marvel at the beauty of Saturn's rings, or see the cosmic star-birth nursery M42 in Orion's sword. Or turn my telescope, fitted with a solar filter, toward the Sun to watch a partial solar eclipse, as I did in October 2014, when to my astonishment I saw a giant sunspot complex larger than any I'd ever witnessed. It turned out to be AR 2192, the largest sunspot group in more than two decades. The next few days, it was in the news, but I got great satisfaction from having seen it before that.

Two other examples: While trekking in East Africa, with the help of our Tanzanian-born American guide, we saw examples of evolution in action in myriad ways among the animals and plants large and small. And at Olduvai Gorge, an African guide noted proudly to us that this gorge, and other sites like it in that part of Africa, are where all mankind evolved. That is thrilling.

Personal observation isn't even necessary. We are so lucky to live in this time. We have our visual exploratory surrogates. Think of the NASA Cassini spacecraft's outside-in view of the solar system (taken July 19, 2013) displaying a gorgeous backlit Saturn, its rings in the foreground, with seven of its moons. At the lower right, inside Saturn's outermost E Ring and far, far behind, was a barely visible blue dot—Earth from 898 million miles away.[1] It's an updated echo of that earlier famous photo taken by Voyager 1 in 1990 from beyond the orbits of Neptune and Pluto, of what Carl Sagan poetically christened the "pale blue dot," a few-pixels-wide image of Earth from 3.7 billion miles away.

Or think of those glorious, first-ever close-up images of Pluto as NASA's New Horizons spacecraft flew by this now reclassified but still beloved distant dwarf planet and its five moons in July 2015. Or the hard landing of the Philae probe from the Rosetta spacecraft onto comet 67P/Churyumov-Gerasimenko. Or the Dawn spacecraft's images of bright patches, thought to be glistening salt deposits, inside craters on the dwarf planet Ceres, which it began orbiting in April 2015, after its earlier mission to the asteroid Vesta.

Consider Cassini's October 2013 image of Saturn from *above*, its rings seen full-on, another view impossible from Earth. Or the NASA spacecraft Juno's spectacularly beautiful images of the chaotic clusters of swirling storms across Jupiter's south pole in 2017 (they look like abstract paintings worthy of hanging in a gallery). Or Apollo 8's Christmas Eve

1968 view of Earth rising from behind a cold and foreboding Moon, an image many credit with stimulating the environmental movement. An even more stunning high-resolution image of the full Earth rising above the gray lunar surface was taken by NASA's Reconnaissance orbiter in October 2015 and issued by NASA on December 18, 2015.[2] It is spectacular! I doubt anyone could gaze upon this image of our brilliant blue planet and not be moved.

Think of any of the myriad dramatic Hubble Space Telescope photos. One I especially love is an extraordinary deep-field view showing a portion of the sky no bigger in diameter than what you can see through a soda straw. In that amazing, narrow view—just that tiny part of the sky!—you can see thousands of galaxies, extending all the way back to near the birth of the universe. The new and more powerful James Webb Space Telescope offered its first photos of the universe in July 2022. For years to come, we will be reacting in awe to its new insights into our galaxy and far beyond.

I want people to have some sense of their place in the universe and feel part of something wonderful and awesome and awe inspiring. Real science does that. Pseudoscience at first seems to offer such marvels, but it is all a sham. Its marvels are untrue, fictions posing as facts. Pseudoscience diverts our attention down meaningless, dead-end, scientifically sterile paths. It wastes time, money, and human capital. And yet—and yet—the scientific study of pseudoscience is a fascinating area of inquiry of its own. It is important, as well. The study of pseudoscience has attracted some of the world's best scientists and philosophers. Embodied in the subject are many surprisingly profound and nuanced intellectual issues. The topic isn't so cut and dried as one might think. There is payoff in devoting attention to it. Some scientists and teachers even use the innate popular interest in pseudoscience as a teaching tool to gain students' interest and to get them to start thinking about how reliable some of the ideas and claims they hear might be.

What to do about pseudoscience? How best to help others see the differences between it and real science? What do those differences tell us about how science is done and how pseudosciences operate? How much does it all matter? These and dozens of other questions have also attracted legions of scientists, scholars, teachers, writers, and investigators. There are also vast numbers of other intelligent people who just like and appreciate genuine science and want their friends and fellow citizens to benefit from more exposure to it, uncontaminated by spurious science. This is especially important for the next generations, the young people who will be making all the decisions in the future.

In this book, I explore the topic of pseudoscience in all its myriad manifestations. What is it? How can we identify it? Why is it so endemic in what we might consider a scientifically enlightened age? I consider it as a persistent irritant to science and an impediment to our understanding of real science, to be sure. But, as I've hinted, I get into deeper, even more interesting issues and questions about it. What is the proper approach of concerned, scientifically oriented people to these shadowy pretend sciences that fleetingly come and go but never quite entirely disappear? What tone should we adopt in talking about it with the public?

I give scores of examples and try to find some organizational structure to the whole set of subjects. And although, like science, pseudoscience is universal in its appeal, I look at a few of the specific nationalistic fervors of pseudoscience that infect specific countries and regions. I use climate science as an extended case study for antiscience, showing the forces (borrowed from the powerful antitobacco lobby of earlier decades) that actively fought climate science and its public acceptance for decades to protect fossil fuels and other vested economic interests. The other antisciences (antievolution, antivaccines, anti-GMOs, and all the others) have their own stories, but they all show many similarities to the climate situation.

To examine pseudoscience and inform the public about it, I describe the rise of an organized movement of scientific skeptics, first in the United States and then globally, and I look at some of the great investigations of pseudoscientific claims. I think you'll see why the subject is so perennially fascinating. You'll see why it remains socially and culturally significant and still poses so many challenges to science, public education, and public discourse, even to our democracy—perhaps especially to our democracy.

I strive to be as generous as possible to innocent consumers of pseudoscience, people fascinated by its ready appeal but unknowingly deceived by its methods and falsehoods. They are its unwitting victims. They have fallen prey to its appeal and its fascinations. I do not criticize them for their interest. The topics are inherently interesting, even fascinating, and I can sympathize with anyone who gets into them for that reason. But we all need better tools of critical thinking to find our way back out of the morass. No information comes at us with the label "pseudoscience" on its front. It is never that easy. Pseudoscience comes clothed as science, just like all the real scientific discoveries we'll read about in future textbooks.

And that's the problem. I hope my discussion can honestly help those innocently deceived. Not one of us, after all, is immune from the ready appeal of pitches that resound to our deepest desires and yearnings. We wouldn't be human if we didn't find these appeals at least initially inviting. We can all sympathize with our friends who get caught up in the tempting pitches. Many other people are victims of what can probably be called self-delusion. They, of course, are complicit in that to some degree; but, not realizing that, they still merit our sympathy and concern. I reserve my criticism, my scorn, for those who *intentionally* deceive. Their numbers are legion.

This isn't a debunking book per se, although you'll read about some major debunkings that resulted from open-minded investigation. It's not in itself a tutorial on critical thinking or on cognitive biases, although the tools of critical thinking and an understanding of our own biased thinking are key ingredients. Our investigative and evaluative examples—our stories of how scientists and other inquirers successfully examined pseudoscientific claims—will, I think you will see, vividly show those fine attributes in action.

This book is, at heart, an inquiry into how to think about pseudoscience. What is the proper approach and tone to take? Should we even use the term *pseudoscience*? If not, is there a better one? Do scientists and investigators sometimes overreach in their rhetoric and opposition to pseudoscience? What is the best way to discuss these issues with the public? I look at tricky semantic issues that befuddle the topic. And I try to place the whole complex subject of pseudoscience into some proper perspective—scientific, historical, social, cultural. The human stories of investigations, past and present, highlight how scientists and others have struggled to apply the tools of science and reason to get to the truth of extraordinary claims. And because all these activities are carried out by human beings, we won't expect perfection. All of us can be, and certainly are, wrong at times. But the processes of science help winnow out the human deficiencies within us all and arrive somewhat closer to the truth.

As a dark shadow of science, pseudoscience will always be with us. It will never be shaken. Pseudoscience has too many glossy attributes that appeal mightily to everyone. It is good at telling people what they want to hear. It has many fans and supporters (even if often they don't realize they are supporting pseudoscientific practices). Pseudoscience has proved to be enormously resilient, flexible, and adaptable. You might even say it evolves. So it is best that we encounter it head on, with our eyes and minds open, our brains on high alert, but our hearts also engaged. We can understand the ready appeal of pseudoscience and be sympathetic to its innocent consumers while realizing that real science always has and always will, in the end, prevail—even if not without difficulty. Science, after all, has the ultimate advantage: It has on its side reality and truth.

CHAPTER THREE
WHAT THE HECK *IS* PSEUDOSCIENCE?

Not long ago, a reader forwarded to me a webpage touting a new product that can presumably detect and diagnose diseases. It was for something called a quantum magnetic resonance analyzer, specifically model AH-Q8. My correspondent found it literally unbelievable. After a quick reading I had to agree. Here is some of what it said:

> The electromagnetic wave signals emitted by the human body represent the specific state of the human body, and the emitted electromagnetic wave signals are different under different conditions of the human body, such as health, sub-health, disease, etc. If we can determine these specific electromagnetic wave signals, we can determine the status of the body's life.
>
> Quantum medicine considers that the most fundamental reason for falling sick is that the spin of electrons outside the atomic nucleus and the orbit change, thereby causing the change of atoms constituting a material, the change of small biomolecules, the change of big biomolecules, the change of all the cells and finally the change of organs. . . .
>
> The frequency and energy of the weak magnetic field of hair determined directly or by holding a sensor by hand compared with the resonance spectra of standard quantum of diseases and nutrition indicators set in the instrument after the frequency and energy are amplified by the instrument and processed by the computer, and then the corresponding quantum value being from negative to positive is output. The size of the quantum value indicates the nature and extent of the disease and the nutrition levels.[1]

Apart from the obvious shortcomings of grammar and syntax, this passage seems mostly gobbledygook. This product description tosses around real scientific terms—*quantum*, *atomic nucleus*, *spin of electrons*, *orbit changes*, *electromagnetic waves*. It then adds questionable concepts—quantum medicine, the assertion that disease and nutrition change atoms' orbital spins—and throws all of it onto the wall, where it sticks in more or less random, meaningless order.

My answer to our correspondent was that this description was the greatest amount of pseudoscientific gibberish in the shortest amount of space I had seen in quite some time. Yet, apart from the density of nonsense, it is not all that unusual.

What Is Pseudoscience?

What exactly is pseudoscience? *Merriam-Webster's New Collegiate Dictionary* definition is a "system of theories, assumptions, and methods erroneously regarded as scientific." That is pretty good. That definition is about as succinct as the matter can be put. The *Oxford English Dictionary* (OED) calls pseudoscience a "pretended or spurious science; a collection of related beliefs about the world mistakenly regarded as being based on scientific method or as having the status that scientific truths might have." I like that definition even better.

Here is my attempt at a definition. It is fairly close to OED's in length and concept: "A set of strong ideas and assertions about the natural world unsupported or even contradicted by the best available scientific evidence and advanced as science without applying scientific methodologies." Note that in all these definitions, the key point is not that the science put forth is wrong. It is that it is *not science* at all. The key phrases are *erroneously regarded as scientific* or *a pretended or spurious science* or, as I put it, *without applying scientific methodologies*.

Merriam-Webster indicates that the word *pseudoscience* first came into use in 1844, but recently, Daniel Thurs and Ronald Numbers found *pseudo-science* used in 1824 in an oblique reference to phrenology. Still, the term remained relatively rare until later in the nineteenth century. They suggest this circumstance is due to the fact that *science* itself was still a somewhat amorphous term.[2]

As I mention in chapter 1, new science at the frontiers is always provisional and will frequently turn out to be wrong. Much of it gets modified or superseded fairly quickly. If you listen to scientific presentations or read scientific papers, as I have all my life, then you will find that the authors almost always acknowledge shortcomings of their study. They point out the need for more observations, larger test samples, better instrumentation, and possible improvements in experimental design and their hope for more insightful theory or interpretations. No single paper is ever the be-all and end-all. No one expects it to be. But usually there is a framework on which future studies can build. Science proceeds that way, usually slowly.

So one key point about a pseudoscientific claim to knowledge is that not only is it wrong (or very likely so) but also that it wasn't achieved through any of the normal methods of science that help provide some minimal threshold of evidentiary quality. In fact, in pseudoscientific activities, methods of real scientific inquiry are typically not only *not applied* but also often not even thought necessary or important. In worst-case situations, they are not even known or understood. (In a visit to China to examine people who claimed to have special powers, we encountered a woman, a police officer, who said she could see inside another person's body and diagnose disease. We asked her if she had ever subjected her ability to a controlled experiment; she had no idea what we were talking about.)

Nevertheless there is a pretension to science, even if not stated. Of course, most respected and legitimate fields of human endeavor are not science (art, music, literature, sports, entertainment, to give just five examples), but they make no such claim. They have their own standards of quality and creativity. It is the implied pretension to being scientific without being so that makes a field vulnerable to being labeled pseudoscience. Let's get to some examples.

Here are five fields where pseudoscientific claims are especially rampant: astrology, ufology, psychic abilities, cryptozoology, and alternative medicine (there are many more). All make big scientific claims—often without even realizing so. Astrology implies an unknown set of forces that affect human beings in different ways depending on the positions of the planets and the date of a person's birth. Ufology takes almost for granted the reality of the two very things that are still in question: the discovery of extraterrestrial beings and the assumption that they routinely visit Earth in spaceships. So-called psychics assume the existence of several strange mental forces or powers unknown to science that can be sensed or communicated across distances (ESP, clairvoyance, psychokinesis). Cryptozoology has its own spectrum of pseudoscience to science, but at the pseudoscientific end, the existence of large creatures (bipedal primates and lake and sea monsters unknown to science) is a given, and the goal is merely to search for evidence to persuade others.

Alternative medicine is such a huge field that it requires a book of its own, but its contentions are rife with dubious and pseudoscientific ideas. Among them are "energy fields" unknown to science, supposed ancient meridians in the body that correspond to nothing in modern human anatomy, the belief that anything natural is superior, and the incredible idea that diluting a solution to the point that no molecule of the original substance can remain is still somehow effective (homeopathy).

Each of these implied assertions is a strong scientific claim. One might even say an extraordinary claim. The discovery of extraterrestrial intelligent life would be arguably the greatest discovery in the history of humanity. Using powerful radio telescopes, astronomers and other scientists have been actively engaged in the search for extraterrestrial intelligence (SETI) for more than a half-century now, so far without success. It is a low-cost, high-risk, high-payoff scientific activity. (The high-risk part—by this I mean the low probability of success—is one reason but not the only one that the federal government stopped funding SETI programs in the 1990s. Other reasons involved some regrettable confusion between real science and pseudoscience. I get to those later.) Scientists would love to find evidence of signals from extraterrestrials (ETs). It would be a revolutionary discovery. Nobel Prizes might follow. The implications for science, technology, philosophy, theology, religion, and all of human culture would be epic. I first wrote about some of these possible implications in an invited chapter in the book *Extraterrestrial Intelligence: The First Encounter* published in 1976. Among the book's other, better-known contributors were science-fiction greats Ray Bradbury and Isaac Asimov, astronomer George Abell, theologian/philosopher William Hamilton, and *Star Trek* actor Leonard Nimoy (Spock).

I wrote my chapter while I was editor of *Science News*. One point I made is worth repeating here: Despite the shock that the discovery of extraterrestrial intelligence might be to society, it should be announced to the public as soon as confirmed. I offer no support to those who suggest the news should be withheld because we are too weak or immature to absorb the information. In that chapter, I conclude, "Facts are facts and truth is truth, and hiding them or hiding from them is futile."[3] History is replete with examples of the damage caused by trying to withhold important information from the public. I mention this here only because in the past decade I have been accused by one of the more extreme and vociferous UFO believers of being involved in a government cover-up of knowledge of UFOs. I am not. I am proud to have been on public record for nearly five decades championing open disclosure if ever and whenever a discovery of extraterrestrial intelligence might be made.

Extraterrestrial Intelligence is a speculative, philosophical, "what if?" kind of work. What *if* we contact intelligent extraterrestrial beings? How would it affect us? At the time, we authors were fairly optimistic (perhaps overoptimistic) that we might make such contact in the coming decades. So far that hasn't happened. It is looking more and more likely that it might not happen for a long time to come. Yet almost all the people who write credulously about UFOs and alien visitations casually assume the discovery has already been made! That is astonishing to me and to most scientists. That is pseudoscientific thinking.

Astrology is an ancient art and belief system that provides comfort to its followers. It may have seemed to make some sense in ancient times, before modern science discovered the size and extent of the solar system, the nature of the planets, the laws of gravity, and other laws of nature. But it prevails, still with ardent followers, in the twenty-first century. The National Science Board's 2014 *Science Indicators* report found surprisingly that Americans were more favorable to astrology in 2012 than at any time since 1983. Only about half said astrology is "not at all scientific"; the rest thought it was either "sort of scientific" (32 percent) or "very scientific" (10 percent).[4]

No one can or even tries to explain why the relative position of the distant planets should affect our lives personally. Proponents offer no satisfactory evidence and no theory. (Most horoscopes are general enough to seem true to nearly everyone.) If astrology were true, then we would have been able rather easily to demonstrate that fact scientifically and produce some good scientific theory explaining the connections. That hasn't happened. And that goes not for just newspaper horoscopes but also the so-called more intricate ancient and modern astrological systems (of which there are many, all different). Given what we know now, the ideas of astrology, attractive and charming as they may seem, are both scientifically dubious and supremely egocentric. Yet in some circles, the confusion between astrology and the real science of astronomy is still with us. Today, professional astrologers use computers and still apply scientific-sounding lingo, but there is nothing scientific at all about what they do.

"Psychics" come in all varieties. (The use of quotation marks to enclose the word here is intentional; it indicates that although people may call themselves psychics, it doesn't mean they possess psychic powers, which is the claim in contention.) They believe or pretend to believe that they have special powers. Uri Geller initially fooled some very good physicists into believing he was bending spoons with the power of his mind. Some were so amazed that for a time they considered this to be real scientific evidence of psychokinesis (PK). That is, until magician James Randi and other skeptical investigators came along and revealed how he was doing it. Geller was surreptitiously bending the spoons mechanically while our attention was distracted elsewhere and then revealing the bend in a way that seemed it was happening before our eyes. There are countless ways to do that. If you prevent one, a magician can easily switch to another. Geller was using common methods of conjuring, yet he didn't bend spoons in the spirit of a magician's performance. He presented it as scientifically real. It was not.

Countless psychics claim to be able to use mental powers to see things that other people cannot, to foretell future events, to find lost bodies, and to solve crimes. None of these claims have ever remotely been validated. Instead, these so-called psychics use a variety of well-known techniques to *persuade us* and others that they have been successful. The alleged powers that lie behind their claims—if they existed and if they had them—would be an astonishing scientific discovery. It would change our world; everything from economics to stock markets to insurance actuary tables to gambling casinos would be much different. Despite persistent assertions to the contrary, no good evidence for them has been found.[5] When Sylvia Browne died in late 2013, she was perhaps the most popular so-called psychic in America. She was a fixture on the *Montel Williams Show*. She made enormous sums from her personal appearances. She had claimed astonishing successes. But multiple independent investigations showed no such success. One detailed study found that of all 116 cases in which she had made a public prediction in a missing person or death, her success rate was zero. She had not once, even remotely, been mostly right.[6] That's hardly what one would assume from the rhetoric of Browne and her promoters and fans. They were engaged in pseudoscientific thinking.

Parapsychology—the laboratory study of such claims of alleged psychic powers—is another field that exists along a spectrum. At one end is reasonably legitimate (but not yet generally accepted) science. At this more scientific end of the spectrum, parapsychologists are trained scientists and try to be careful and responsible experimenters and observers. But even their work has not persuaded most psychologists and other scientists that any special powers or forces unknown to science have been demonstrated.[7] The assertion of psychic powers as a reality without acknowledging this lack of agreed-upon positive scientific evidence—and also the huge number of *negative* studies—is pseudoscientific thinking.

So pseudoscience is all around us. It doesn't call itself that. Most people never use the term. You don't hear it applied very often. That's why, unless you have been following these issues from a scientific viewpoint for a long time, some of what I've said so far may surprise you. But assertions of a strong scientific claim in the absence of any equivalently strong scientific evidence—and especially when ignoring strong evidence against it—is pseudoscience, whether its practitioners think of it that way or not.

CHAPTER FOUR
WHAT'S THE HARM? WHY DOES IT MATTER?

Well, what's the problem? What does it matter? Who cares? I can almost hear the rumbles of questions along these lines. That's good. Questioning is good. In fact, it is part of the process of scientific skepticism and one of the important characteristics of science itself. Ask important questions. Don't take anything at face value.

Do you think it's important for significant decisions that affect you to be based on reasonably accurate information? Most people do. When you go to buy a car, you don't want the salesperson to give you such dubious claims as "It gets one hundred miles per gallon." If he or she starts saying such things (unless it is electric!), then you'd probably ask some penetrating questions about that. You'd question his or her veracity—or just walk away. When your elected officials are voting on a law or legislation that affects your well-being, don't you want them to understand it, at least on some basic, rudimentary level?

We all know people who live in a kind of little fantasy world of their own creation. They may believe in all manner of weird things. In general, that may not a problem for the rest of us. Who's to say what's weird anyway? But most people compartmentalize that bubble of fantasy to just those certain beliefs. When they use the navigation software on their car or their smartphone or use the internet or get on an airplane or visit their primary care physician and get test results or medication, they are using the fruits of modern science and technology. They are in a sense acknowledging that there's a big, real world where they depend on engineered devices and products and discoveries based on the laws of physics and real scientific understanding, where their fantasy beliefs don't apply.

Pseudoscientific thinking exists, as I've said, on a spectrum, from relatively harmless (even goofily charming) to deeply pernicious and dangerous. The American actress Shirley MacLaine is well known for her interests in all sorts of New Age hooey (she might call it alternative spirituality), including channeling ancient entities. She's quite open about that. A few years back, when she went to sell one of her properties in northern New Mexico, a 7,450-acre ranch near Abiquiu, she consulted psychics to arrive at the price. They advised her to sell it for $30 million. Instead, "she settled on $18 million because 'nine is the number of completion' and the digits in 18 (1 and 8) add up to nine."[1]

It may not matter much if your friend wants to believe that the morning's daily horoscope is an accurate guide to their life or that they can channel ancient goddesses or that UFO reports really do represent alien spaceships in our sky. I don't care myself. But problems begin if your friend starts influencing lots of other people. And this is especially the case if the influence extends beyond just their personal belief into an assertion that what they believe in so strongly is empirically, really, actually true—in the sense that gravity is real or the stars are really out there (or at least were when their light left them) or that in the presence of sunlight plants convert carbon dioxide and water into the organic compounds they need for nourishment.

And then what if your friend joins the city council or the county commission or the local school board and starts asserting that the world is only 10,000 years old and that evolution and the age of the Earth and universe and evidence of the big bang should not be part of public schools' science curricula? Now we have serious problems. And in the United States at least, what I describe here is not an uncommon scenario. On almost every local school board and many regional education boards, the voters have elected members who not only hold these pseudoscientific (and antiscientific) views but also would dearly like to insert them into the public policy agenda. Organizations that push these belief systems actively encourage their members to get involved in local political affairs solely so they can advance those views. The members proceed to do just that, sometimes surreptitiously, often openly.

Then think about the same people advancing to the national political stage, say, getting elected to Congress, where they can steer national legislation to support pseudoscientific or antiscience policies. That, sad to say, has happened—repeatedly.

One of our greatest writers on pseudoscientific beliefs, the late Martin Gardner, wrote about this matter way back in 1952 in a book, *In the Name of Science*. Republished in 1957 under the title *Fads and Fallacies in the Name of Science*, the book has influenced generations of scientists and scientific-minded skeptical investigators ever since. (It is still the recommended starting point for anyone wanting to understand the issues of pseudoscience and bogus science.) "If the people want to shell out cash for . . . flummery, what difference does it make?" Gardner asks. "The answer is that it is not amusing at all when people are misled by scientific claptrap."[2]

He goes on to cite suicides and mental crack-ups from people who read crank books rather than get trained psychiatric care, teachers who are afraid to teach evolution because they may lose their jobs, and flying saucer books that sow distrust by blaming military and government leaders for nonexistent cover-ups. Gardner adds,

> An even more regrettable effect produced by the publication of scientific rubbish is the confusion they sow in the minds of gullible readers about what is and what isn't scientific knowledge. And the more the public is confused, the easier it falls prey to doctrines of pseudo-science which may at some future date receive the backing of politically powerful groups.[3]

Gardner was writing here not long after the end of World War II, and the horrors of Nazi Germany were still recent and raw. He explains, "A renaissance of German quasi-science paralleled the rise of Hitler. If the German people had been better trained to distinguish good from bad science, would they have swallowed so easily the insane racial theories of the Nazi anthropologists?"[4] He then went on to make a succinct case for an informed public's need to identify pseudoscience: "In the last analysis, the best means of combating the spread of pseudo-science is an enlightened public, able to distinguish the work of a reputable investigator from the work of the incompetent and self-deluded."[5]

Decades later, Carl Sagan, the noted astronomer and champion science popularizer, eloquently expresses these kinds of concerns in vivid, memorable prose: "We have a civilization based on science and technology, and we've cleverly arranged things so that almost nobody understands science and technology. That is as clear a prescription for disaster as you can imagine."[6] He precedes that statement with a stirring passage about science and democracy:

> It is a foreboding I have—maybe ill-placed—of an America in my children's generation, or my grandchildren's generation, when all the manufacturing industries have slipped away to other countries; when we're a service and information-processing economy; when awesome technological powers are in the hands of a very few, and no one representing the public interest even grasps the issues; when the people (by "the people" I mean the broad population in a democracy) have lost the ability to set their own agendas, or even to knowledgeably question those who do set the agendas; where there is no practice in questioning those in authority; when, clutching our crystals and religiously consulting our horoscopes, our critical faculties in steep decline, unable to distinguish between what's true and what feels good, we slide, almost without noticing, into superstition and darkness.[7]

Sagan spoke these words, with their powerful final lines, in June 1994 as keynote speaker before a large audience at the conference of our Committee for the Scientific Investigation of Claims of the Paranormal (CSICOP) in Seattle. It had been my honor to introduce him. They were first published under the title "Wonder and Skepticism" in our magazine the *Skeptical Inquirer* in January/February 1995. He expresses similar thoughts, with greater elaboration, in his foundational 1996 book *The Demon-Haunted World: Science as a Candle in the Dark.* The book was published shortly before his untimely death from myelodysplasia, a bone marrow disease, in Seattle on December 20, 1996. It is an important book, some think his most significant, but Sagan was too ill to promote it in the way he did his earlier books. Whether the forebodings he expresses in "Wonder and Skepticism" would have diminished or increased had he lived until now, I leave to you to consider.

Citizenship in a Democracy

The biggest worry about pseudoscience by people who like and respect science—myself included—is the problem of accurately informing the citizenship in a democracy. Almost all scientific leaders speak of the need for the voters and the people they elect to have at least some basic level of scientific literacy. By some counts, more than half of all public policy issues now involve science and technology in some way. The issues are difficult enough on their own, but it is surprising how often pseudoscientific or antiscientific thinking creeps into politicians' thinking (often at the behest of constituents). Not knowing some rudimentary scientific principles, not having a basic level of scientific knowledge, makes them exceedingly vulnerable to such proposals. That situation has greatly worsened in the last few years, with an increasing number of elected legislators evoking clearly antiscientific and pseudoscientific concepts. This is evident in their views and statements about climate science, the efficacy of COVID vaccinations, and much more.

Another great popularizer of science, astrophysicist Neil deGrasse Tyson, the Hayden Planetarium astrophysicist and narrator of the new *Cosmos* series (2014, 2020; continuing Sagan's), has spoken about these issues. He encourages scientific literacy *before* people reach adulthood and get into positions of influence because otherwise it may be too late; they won't have a pseudoscience detector:

> True science literacy basically reveals the folly of so much of what's going on in the pseudoscientific world. I'm really out to support science literacy in the nation as being good for a democratic society. . . . If you have a scientifically illiterate public, they can't possibly be in control of their future, given that in the twenty-first century the role of science and technology will be greater than ever before. . . . I'd rather get people thinking straight in the first place, rather than this huge effort of undoing tangled mental pathways that occurs later on in adulthood.[8]

"I invest considerable time in how people think," Tyson said in a 2021 interview for our *Skeptical Inquirer Presents* online webinar. "What receptors they have for learning. If you go 90 percent of the way to them, you might have a chance. . . . We're all on the same side here. We all want to move into a more rational world. It's an eternal battle but a battle worth fighting."[9]

Consumer Protection

Another strong motivation for examining pseudoscientific claims is consumer protection. Pseudoscientific practices often have acute personal effects. Examining and exposing them is a public service. Psychics come out of the woodwork when people are searching for missing loved ones. Sometimes their efforts are invited, but often they aren't. What they say can be devastating.

Many years ago, there was a famous incident in the high mountains of the Andes. A plane on its way from Montevideo, Uruguay, to Santiago, Chile, carrying forty people, including Uruguay's entire fifteen-member rugby team and a number of their friends and relatives, disappeared. It was early spring, and there had been heavy winter snows. Searches continued for weeks, to no avail. A psychic named Gerard Croiset told authorities to search around one particular set of mountains and landscape, which they did, devoting considerable resources to that effort and location, forty-one miles south of a place called Planchon. Later, it turned out that the plane had crashed at an entirely different location, forty-one miles *north* of Planchon. Eighty-two miles in high, mountainous country is a world of distance away. The plane was not under a mountain nor in or near a lake as he had said. Only sixteen of the people aboard survived the terrible ten-week ordeal. At least some of the parents felt that Croiset had set searchers off on a false trail that greatly delayed finding their loved ones who had still been alive at the time.[10]

Jumping ahead to recent times, the Amanda Berry case presents a distressing example of just how wrong a psychic can be—and how a psychic's false statement can have devastating personal effects. Amanda was one day short of her seventeenth birthday on April 21, 2003, when she went missing from her home in Cleveland, Ohio. No trace could be found. The next year her mother, Louwana Miller, went on the *Montel Williams Show* with Sylvia Browne. Browne told her, "She's not alive, honey." The mother was devastated. Grief-stricken, she died in 2006; some reports said of heart failure. Then on May 6, 2013, Amanda Berry and two other abducted women, Gina DeJesus (missing since 2004) and Michelle Knight (kidnapped in 2000), escaped from a house where they had been held captive by a local man, Ariel Castro. They were survivors of a terrible ordeal, but Amanda's mother was not there to welcome her.

This wasn't even the first time Browne had done something like this. In 2002, also on the *Montel Williams Show*, she told the parents of missing child Shawn Hornbeck that their son was dead and his body would be found in a wooded area near two large boulders. Five years later, on January 16, 2007, Shawn and another boy were found alive in the home of a man named Michael Devlin. He had kidnapped them both. Browne's predictions were wrong in every detail—especially in saying Shawn was dead.[11]

Of course, the more general issue is why does anyone believe what a psychic says anyway? The reasons are legion, and they go to the heart of human psychology. But having a questioning attitude and a skeptical, scientific perspective about such things is one of the best antidotes to being deceived. Claims of psychic powers have been tested so many times and, as a result, debunked so persuasively to those with scientific understanding that it does take a kind of magical, pseudoscientific thinking to give them much credence.[12]

In 2019, the story emerged of a woman yoga instructor who trusted her intuition so strongly that it almost cost her her life after she became lost in a forest preserve in Hawaii. Amanda Eller went on a short run without her cell phone or water and wasn't found for seventeen days. She survived on berries and river water. My colleague Ben Radford looked into her case and showed that it was an overreliance on what she called intuition that caused her to inadvertently go deeper into the forest instead of in the direction of her car. She said in an interview, "I have a strong sense of internal guidance, whatever you want to call that, a voice, spirit, heart. . . . My heart was telling me, 'Walk down this path go left. . . . Great! Go right. This is so strong! Go left, go right."[13] She later said in a video posted to her Facebook page that after getting lost and meditating, she spent a couple hours searching: "I felt that was the direction of my car. Clearly it was the wrong way, and I continued to go that way. . . . The only thing I have is my gut; I don't have a compass, I don't have a cell phone, so [I said to] the spirit or whatever you want to pray to, 'Please help guide me.' . . . That's what led me five miles away [from safety]."[14] She told the *New York Times* she was also into numerology, so every day she was thinking of the date and looking to some link that would be her lucky day. She called the whole thing a "spiritual" experience. Her reliance on her gut took her the wrong way, almost cost her own life, and put search parties in danger as well.

There are so many stories we hear in which one's intuition supposedly is responsible for guiding a person to a right decision; we seldom hear the stories where intuition had the opposite effect. Radford points out that this effect and confirmation bias tend to make us think our intuition is more correct than it actually is. This is not a knock on intuition, but "psychics" often rely on something very similar, and this is how they can deceive themselves into thinking they have special powers. It all is a cautionary note that nonrational thinking can lead us into danger.[15]

Quackery and Alternative Medicine

What most people now generously call "alternative medicine" is another area where great harm can come. Quackery and the marketing of questionable medical remedies is big business. How big? Hard to say, for sure. But it is *big*.

In the 1980s, there actually was a major congressional investigation of quackery and its costs. This, of course, happened before Congress came under the sway of a few influential politicians sympathetic to untested medical remedies, which led to the establishment in 1998 of the National Center for Alternative and Complementary Medicine (NCACM; in December 2014, NCACM changed its name to the National Center for Complementary and Integrative Health, most likely because so many "alternative medicines" have proved ineffective). But in the early 1980s, the Select Committee on Aging's Subcommittee on Health and Long-Term Care, under the chairmanship of Senator Claude Pepper of Florida, conducted an intensive four-year investigation of quackery and its impact on the elderly. The investigation was the most comprehensive investigation of medical quackery and related health care fraud ever undertaken, according to its 1984 report *Quackery: A $10 Billion Scandal*.[16] It grew out of a series of hearings that had identified health fraud as the single most prevalent and damaging of the frauds directed at the elderly. The committee's report uses the term *quack* for "anyone who promotes medical schemes or remedies known to be false, or which are unproven, for a profit" and *quackery* as the "practice or pretensions of a quack."[17]

The subcommittee estimates the "cost of quackery—the promotion and sale of useless remedies promising relief from chronic and critical health conditions—exceeds $10 billion a year." It adds, "The cost of quackery in human terms, measured in disillusion, pain, relief forsaken or postponed because of reliance on unproven methods, is more difficult to measure, but nonetheless is real. All too frequently, the purchaser has paid with his life."[18] Phony cancer cures, questionable arthritis cures, and phony antiaging remedies (the fastest-growing problem) made up the bulk of the quack treatments at that time.[19] The 250-page report laments that while the impact of quackery on our lives has been increasing and growing in sophistication, public and private efforts to control the problem "have diminished, been redirected, or disbanded." It says the US Food and Drug Administration, "once a formidable force in controlling quackery, now directs less than 0.001 percent of its budget to controlling quackery."[20]

The report doesn't take on all unproven remedies. Pepper and the Senate investigators realized, correctly, that the "healing art continues to evolve" and "some of what is unproven may yet prove of benefit." Instead, it focuses on a specific set of unproven remedies. The common denominator is "conscious deceit and the absence of, and, in most cases, total disregard for, scientific proof. Quackery represents pseudoscience at best."[21] Things are pretty bad when the *best* thing you can say for some remedy is that it is pseudoscience.

The report outlines a number of fallacious or questionable health beliefs people hold; for example, extra vitamins provide more pep and energy, and arthritis and cancer are caused by vitamin and mineral deficiencies. Another is that most people believed, wrongly, that advertisers are so rigorously policed and regulated that serious distortions are very unlikely or improbable. The report says,

> On these beliefs, needs, misunderstandings and fears, the quacks build and prosper. The purveyors of quackery provide a panacea for virtually all human ailments. They are unconfined by the need to be credentialed and establish scientific proof. . . . Beneath their mask of respectability lies more harm than just the dollars diverted from those in need. Their take includes the health that could be protected and improved by proper medical procedures. It must also be measured in terms of disillusion, despair, misery, and death.[22]

Note that this investigation took place just before the modern rise of alternative and complementary medicine and its powerful supporters in mainstream media and public policy. It was also before the rise of the internet and cable TV channels with their ubiquitous infomercials and other solicitations of dubious products and remedies. So the situation is, if anything, very likely worse today. And of course, the $10 billion-a-year estimate (1984 costs) must be multiplied by inflation and the population increase since then.

Also notice that the Pepper Committee's definition of *quackery* is narrower than our definitions of *pseudoscience*. *Quackery* is limited to the promotion of remedies unproven or known to be false *for profit* (my emphasis). Much pseudoscience—and there is no reason to think medical pseudosciences are that much different—is carried out by sincere believers. So the problem of pseudoscience in the medical field is a broader and much bigger issue than the problem of quackery itself. Quackery may be the most egregious aspect, but it is not the only one.

Six years after the Pepper study, another study was published, and it got considerable publicity. This one examines the prevalence, costs, and patterns of use of "unconventional medicine" in the United States. Its lead authors are physician David M. Eisenberg of the Harvard Medical School and five colleagues from Harvard, the University of Michigan Institute of Social Research, and several major hospitals. They conducted a national survey focusing on sixteen interventions generally considered to be "unconventional therapies." They define *unconventional therapies* as "medical practices not taught widely at U.S. medical schools or generally available at U.S. hospitals."[23] Prime examples include acupuncture, chiropractic, and massage therapy. Other examples are relaxation techniques, imagery, spiritual healing, commercial weight-loss programs, lifestyle diets (e.g., macrobiotics), herbal medicine, megavitamin therapy, self-help groups, energy healing, biofeedback, hypnosis, and homeopathy.

One in three respondents reported using at least one unconventional therapy in the previous year, and a third of these went to providers for unconventional therapy. Most visits, they found, were for chronic, not life-threatening, medical conditions. Extrapolated to the US population, their study suggests that in one year, Americans made an estimated 425 million visits to providers of unconventional therapy, exceeding the number of visits to US primary care physicians (388 million). Expenditures for these unconventional treatments totaled $13.7 billion, three-quarters of which was paid out of pocket. This figure is more than was spent that year out of pocket for hospitalizations in the United States ($12.8 billion).[24] Questionable and pseudoscientific medical remedies kill people and make others sick who don't need to be. Bogus cancer remedies attract vulnerable patients, soak up their monetary resources, and distract them from real medical care that might save them.

This study caused some controversy among science-minded people. It helped establish the questionable view that alternative medicine was more mainstream than had been thought. Note that several of the study's "unconventional" categories, including relaxation therapy and self-help groups, are not necessarily that unusual or that threatening to good science and are very widely practiced. These inclusions served to widen the numbers of people said to be seeking "unconventional" practices. Whether that was the intention or not, it led to a more sympathetic view of the field for the next decade or so.

Antivaccination movements have caused a recent emergence of once-controlled diseases and made the COVID-19 pandemic continue with multiple resurgences. Andrew Wakefield's bogus theories and claims about the connection between autism and measles vaccines (later retracted by the medical journal that published them) caused worried parents not to have their children vaccinated, exposing their children (and others) to measles, a disease they needn't get. Measles rates, as a result, are rising alarmingly in many developed countries. Other long-preventable diseases are coming back, as well. Pseudoscientific theories that AIDS is not caused by the HIV virus have had terrible effects in South Africa and other African countries. Thousands have died needlessly. (Medical-related pseudosciences are so common that I devote chapter 6 to them.)

What I describe here is hardly an exhaustive portrayal of the harm pseudoscience can do. These are only a few representative examples. If you want a full report of examples of harm from pseudosciences, consult Tim Farley's website, *What's the Harm?* (whatstheharm.net). It summarizes documented stories of harm to nearly 700,000 people from scores of medical and other pseudosciences (all the kinds of subjects in this book). The site makes a point about the danger of popular topics being promoted via misinformation and not thinking critically about them. The accumulated record of tragedy is numerically totaled at the top this way: "368,379 people killed, 306,096 injured and over $2,815,931,000 in economic damages."[25] The website hasn't updated these numbers in several years, so they are undoubtedly higher today.

People are hardly the only victims of pseudoscience. Just ask the rhinos and elephants of Africa. Pseudoscientific myths and superstitious notions about the medicinal value of wildlife body parts contribute mightily to the destruction of elephants and rhinos in what can accurately and unfortunately be called a wildlife apocalypse. Elephants are killed in massive numbers for their tusks, and rhinos for their horns, solely because traditional Chinese medicine, filled with pseudoscientific and disproved notions, promotes grinding them into a powder for numerous unproven treatments, cures, and aphrodisiacs. It is ludicrous because rhino horns are composed mainly of keratin, the same substance in human hair and fingernails, yet a kilogram of powered rhino horn can sell in China or Southeast Asia for $60,000. The enormous economic demand caused by these pseudoscientific notions is driving the extinction of rhinos and elephants, both majestic creatures whose loss will be a blot on humanity. We can attribute their demise to colossal human arrogance and pseudoscience-fueled ignorance. (My colleague Bob Ladendorf and his brother Brett survey this tragic situation in a *Skeptical Inquirer* cover article, "Wildlife Apocalypse."[26])

Pseudoscience's Effects on Science

Pseudoscience can also have deleterious effects on science itself. Sometimes a whole field of science can be stalled when some related pseudoscientific proposition attains a large following. The field becomes discredited and distorted for years. Perhaps the most famous example is how genetics research in the Soviet Union was set back more than two generations starting in the 1930s by the state-sponsored pseudoscientific theories of Trofim Lysenko.

Lysenko captured Stalin's favor by advocating pseudoscientific ideas about how to increase the yield of wheat crops, reinforcing Marxist doctrine. Scientists with real knowledge of plants, including the courageous plant geneticist Nikolay Vavilov, the primary target of Lysenko's (and thus Stalin's) ire, were suppressed, persecuted, and tortured. Genetics became a taboo topic. The whole affair set back Soviet biology for decades.

Lysenkoism has long been cited as a pernicious example of what can happen when a political regime embraces a pseudoscientific concept. It can seem too real in recent political climates that foster cultural divisions and grievances, including antiscientific and pseudoscientific ideas, for instance, about vaccines or climate science.

Some effects on science might surprise you. In the 1970s, Peter Tomkins and Christopher Bird wrote a book called *The Secret Life of Plants*. It became a phenomenon, making the *New York Times* bestseller list and getting people everywhere talking to their plants. Relying on reports of uncontrolled experiments, random observations, and anecdotes, the authors put together a case that plants can count, communicate with each other, and receive signals from life forms elsewhere in the universe.[27]

One of the more bizarre claims is that exposing photographs of growing plants to particular wavelengths of electromagnetic radiation can rid plants of insect pests or fertilize the soil. Many of the sensational "experiments" described in the book were done by polygraph expert Cleve Backster. Backster hooked houseplants up to his equipment and claimed he found responses equivalent to human emotion. The book was aggressively promoted and became a cultural phenomenon. *New Yorker* cartoons were inspired by the claims. The *Doonesbury* comic strip took note of them. The American Society of Plant Physiologists and the American Association for the Advancement of Science scheduled sessions to evaluate some of the claims. Plant scientists tried to replicate his experiments, without success. Backster was nonplussed.

Another of Backster's claims is that a philodendron plant leaf would respond electrophysiologically to the death of brine shrimp put in scalding water in the plant's presence. In response, three scientists of neurobiology and behavior at the New York State Veterinary College in Ithaca did five controlled experiments to test the claim and published them in the journal *Science*. They found no relationship between brine shrimp killing and the electrical responsiveness of the philodendron.[28] As a result of the embarrassment about the sensational, unsupported claims described in *The Secret Life of Plants*, legitimate scientific research into plant behavior was damaged and held back.

Israeli biologist Daniel Chamovitz, author of the 2012 book *What a Plant Knows*, addresses this issue directly in his prologue. He says

Tompkins and Bird essentially shut down research in this field. He obviously feels that he had to distance himself from them in his own book:

> My book is not *The Secret Life of Plants*; if you're looking for an argument that plants are just like us, you won't find it here. As the renowned plant physiologist Arthur Galston pointed out back in 1974 during the height of interest in this extremely popular but scientifically anemic book, we must be wary of "bizarre claims presented without adequate supporting evidence." Worse than leading the unwary reader astray, *The Secret Life of Plants* led to scientific fallout that stymied important research on plant behavior as scientists became wary of any studies that hinted at parallels between animal senses and plants.[29]

Now plant scientists are beginning to document real effects of responsiveness and networked behavior that some consider analogous to or roughly akin to awareness, communication, information processing, and learning. A vast underground neural network of intricate filaments, called the mycelium, binds the forest together in a collaboration between fungi, plants, bacteria, and animals. In this way, plants can sense light, moisture, pressure, hardness, microbes, and chemical signals from neighboring plants. Through their interconnected systems, trees in a forest can quickly share such information.

There's a new scientific society and a journal, *Plant Signaling and Behavior*, for airing such research. None of it involves contending that plants have telekinetic powers or feel emotions. What's going on with plants and their sensory/communications systems seems to be similar to the kind of emergent behavior we see in networked insect colonies, in which behavior that appears intelligent emerges in the absence of brains or central nervous systems.

Suzanne Simard, a professor of forest ecology at the University of British Columbia, summarizes her groundbreaking research into forest ecosystems in *Finding the Mother Tree: Discovering the Wisdom of the Forest*. Peter Wohlleben's *The Hidden Life of Trees: What They Feel, How They Communicate* and Merlin Sheldrake's *Entangled Life: How Fungi Make Our Worlds, Change Our Minds, and Shape Our Futures* are other excellent and very readable sources.[30] An earlier good discussion of all these new insights and various controversies about them can be found in Michael Pollan's *New Yorker* article "The Intelligent Plant."[31] Ann Druyan dramatized this astonishing new knowledge about forest systems in her 2020 television documentary series (and book of the same title) *Cosmos: Possible Worlds*.[32]

The situation has even been portrayed in fiction. Richard Powers's beautifully written 2018 Pulitzer Prize–winning novel *The Overstory* tells the stories of six sets of characters who love, appreciate, or understand trees and forests in ways most people don't. One character, Patricia Westerford (apparently inspired by Simard), practically grew up in forests because her father, an extension agent, often took her with him and lovingly showed her all he has learned about trees and vegetation: "Plants are willful and crafty."[33]

Her life is a roving tutorial. He teaches her to be superobservant about tree behavior, how to ask questions, how to design experiments. In college, she studies forestry, but she already knows a secret: "Trees are social creatures." She gets a PhD and, as her postdoc, studies sugar maples, soon discovering that when one tree is attacked, nearby trees are somehow alerted and ramp up their defense. She verifies everything: "Her maples are *signaling*. . . . These brainless, stationary trunks are protecting each other." She writes up her conclusions in a carefully written scientific paper that concludes, "The biochemical behavior of trees may make sense only when we see them as members of a community." Her paper gets published, even gets some media attention. But soon, three leading dendrologists, clearly not ready to embrace this new way of thinking about forests and probably influenced by the earlier sensational claims I mention, publish a rebuttal, calling her methods flawed and her statistics problematic. They rip apart her paper. She is devastated.[34]

Deeply discouraged, she drops out of science. Eventually she resumes her investigations and to her great joy is able to show that when underground roots of different Douglas fir trees fuse, the two trees join their vascular systems into one. Her trees, she finds, are networked together underground by thousands of miles of living fungal threads. They all feed and heal each other. They keep their young and sick alive, pool their resources, and are far more social and communal than even she had thought. She is welcomed back to science and becomes an honored figure, but it was a close call. The field had to get used to the idea that, as Powers writes, with a forest, "competition is not separable from endless flavors of cooperation."[35]

The point of all this is that the notoriety of *The Secret Life of Plants* for a long while led to scientists censoring themselves away from exploring these kinds of real systems in plants and the legitimate analogies between animal and plant behavior. Good science was delayed by pseudoscience.

The Velikovsky Affair

I have long felt that the decades-long controversy set off by the publication of Immanuel Velikovsky's pseudoscientific book *Worlds in Collision* in 1950 may have deterred Earth and geophysical scientists from being more open to some *real* externally caused catastrophes in Earth's history. Velikovsky proposes that around 1500 BC, Jupiter ejected a planet, Venus, and had several close encounters (but no collisions) with Mars and Earth starting about five thousand years ago. These encounters included strong electrical and gravitational effects, changed the planets' orbits and spins, and even reversed the direction of Earth's rotation. He asserts, for example, that lunar craters were produced by super-lightning bolts. Even worse, Velikovsky, who was not a scientist but a psychiatrist, relies entirely on ancient myths for his ideas. He feels his ideas explained a set of Old Testament stories—the biblical ten plagues, the parting of the Red Sea, the rain of manna from heaven, and the supposed stopping of the sun in its course during an Israelite battle. In concocting all these imagined astronomical events, Velikovsky ignores the results of archaeology, astronomy, geology, physics, and other sciences. This antiscience bias is a big failing. It keeps him from seeing how wrong his arguments are.

Astronomers and other scientists couldn't understand why Velikovsky's fantastic scenario found so many believers among intelligent people, but that it did. The Velikovsky controversy caused extraordinary suspicion and animosity between scientists and supporters of Velikovsky (including some literary types and other respected opinion leaders), so much so that the feelings spilled over to discredit, for a time at least, the concept of catastrophism in general.

In 2001, noted planetary scientist David Morrison, who over the years had written several key scientific critiques of Velikovsky's astronomical contentions, polled some of his top scientific colleagues about whether the Velikovsky episode had any significant influence on the acceptance of catastrophist ideas in the Earth and planetary sciences over the past half-century, positive or negative. The scientists polled were "among the most creative and even revolutionary researchers in their fields," Morrison says. Twenty-three of the twenty-five polled responded. Nobody said the episode had positive effects. Although fourteen felt there weren't important effects, nine said the effects were negative. Their comments to Morrison gave a sense of their frustration. Morrison summarizes, "The statements of these scientists indicate that none of them saw any value in Velikovsky's theories, and that Velikovsky's reputation sometimes impeded acceptance of their own work, or at least was an irritant when they described their work to the public."[36]

I once did my own poll of some noted planetary scientists about one notorious Velikovsky claim.[37] In April 1980, the year after Velikovsky died, his publisher, Doubleday, placed an ad in the *New York Times Book Review* titled "Velikovsky: The Controversy Continues." It boldly claims that the recent Pioneer spacecraft probe to Venus supports Velikovsky's claims about Venus's origins: "Velikovsky's assertion that Venus is a young planet, expelled from Jupiter only thousands of years ago, has received strong support from evidence of the Pioneer probe."[38]

I could not recall any scientific report indicating that Venus was a very young planet, born in historic times and scooting about in the neighborhood of the Earth as recently as 2,700 years ago. I sent the ad to a number of noted planetary scientists and asked them about its assertions. I received responses from about half and followed up with phone calls to about half of the nonrespondents. They were in overwhelming agreement that the substantive claims in the ad were wrong. "The ad is thoroughly dishonest," said A. G. W. Cameron of the Harvard-Smithsonian Center of Astrophysics. Edward Anders, a University of Chicago geochemist who specializes in chemical studies investigating the early history of the solar system, was even more biting: "The ad, like the Velikovsky books it is promoting, contains more falsehoods in a paragraph than one can refute in a chapter. No reputable scientist 'now speculate[s] that . . . Venus was formed more recently than the other planets.'" Geophysicist William M. Kaula of the University of California at Los Angeles was also blunt: "The statements [in the ad] are not accurate. As a Pioneer Venus investigator, I am not aware of any support in the scientific community for Venus being a young planet." One of the scientists I polled was the aforementioned

David Morrison. Morrison had been a student of Carl Sagan and was a prolific researcher and writer about planetary science missions. He said, "I am used to the distortions by Velikovsky supporters, but this ad seems to be particularly reprehensible."[39]

Assertions like those made in the ad had long contributed to the misimpression that Velikovsky had not been given a fair hearing and was actually onto something. He was not. The planetary scientists I polled were unanimous about that. But I hope you can now see why planetary scientists, like most scientists generally, were so infuriated with Velikovsky's claims and felt that publishers and other literary defenders of his ideas had given them far too much credence. You can see why Velikovsky's aggressively promoted ideas tainted geophysics and impeded the acceptance of some scientifically legitimate catastrophic evidence and insights.

This resistance began to break down in 1980, when University of California geologist Walter Alvarez and his Nobel laureate physicist father, Luis Alvarez, reported strong evidence that the Cretaceous-Tertiary (now called the Cretaceous-Paleogene) extinction event of 65 million years ago was set off by the impact on Earth of an asteroid about 10 kilometers (6 miles) in diameter. A colleague asked me about the claim at a meeting of the American Association for the Advancement of Science in February 1980, and I remember myself—probably still influenced by the Velikovsky debacle—being initially dubious about it.

At that same meeting, Walter Alvarez gave a preliminary report before the paper's publication in the June 6, 1980, issue of the journal *Science*. (The paper had the suitably dramatic title, "Extraterrestrial Cause for the Cretaceous-Tertiary Extinction."[40]) Like most of the scientific community, I soon changed my mind upon seeing the full evidence. The main evidence was the "iridium anomaly," the discovery, at many different locations on Earth, of a greatly elevated abundance of iridium in just that layer of the geological strata that defines the change from the Cretaceous period to the Tertiary, the time of the mass extinctions. The scientists showed why that iridium is likely of extraterrestrial origin (but was not from a nearby supernova, an exploding star). They suggested a collision with a large, Earth-crossing asteroid.

Research since then has vastly strengthened the case for that asteroid impact as the cause of the mass extinctions. The huge 110-mile-wide, 12-mile-deep impact crater it created, named Chicxulub for the Yucatan town near the site's center, has been identified beneath the surface of the Yucatan Peninsula and the adjoining seawaters. The impact had vast global effects on the atmosphere and life. The dinosaurs were the largest of the life forms extinguished over the decades and centuries that followed the impact. Although many geologists and paleontologists took about a decade to accommodate themselves to the idea, the impact hypothesis is now widely accepted by the scientific community. In March 2010, an international panel of forty-one scientists reviewed twenty years of research and endorsed the Chicxulub impact as the cause of the Cretaceous-Paleogene extinction.

But remember Morrison's survey? Here is how Walter Alvarez himself replied to the question about Velikovsky's effect on their science:

> [Velikovsky did not influence science] in any positive ways. I considered him part of the problem we faced in getting a hearing for the KT impact hypothesis, because his ideas, which were incompatible with the laws of physics, had confirmed many geologists in their view that people working on extraterrestrial causes for events in Earth history were not doing good science.[41]

In other words, the legacy of Velikovsky's pseudoscience had impeded initial acceptance of one of the great *real* discoveries about Earth's history.

Is *Pseudoscience* a Pejorative Term?

It is time to face one of the biggest controversies about considering anything pseudoscience. Many sociologists and historians of science and other scholars consider the term pejorative. Some go further to say that using it is unfair, demeaning, and only a way for the scientific community to patrol its borders and keep science "pure." Other scholars, including some prominent philosophers, find value and meaning in the term and support its use. Most scientists just use it or not as they wish, generally ignoring such arguments.

The controversy, I think, is interesting and useful in its own right. It raises many fascinating issues and highlights how different intellectual starting points result in decidedly different views about the perennially troubled borderlands of science. Here's a sample of what I mean.

The late sociologist Marcello Truzzi begins his lengthy entry "Pseudoscience" in *The Encyclopedia of the Paranormal*, "The term 'pseudoscience' along with such other terms as 'pathological science,' 'crank or crackpot science,' 'cargo cult science,' 'garbage or junk science,' 'antiscience,' and 'quackery' (with which it is frequently confused), is a pejorative label often used to denounce claims and methods erroneously represented as scientific."[42]

Philosopher and science critic Larry Laudan states *his* objections even more bluntly: "If we would stand up and be counted on the side of reason, we ought to drop terms like 'pseudo-science' and 'unscientific' from our vocabulary; they are just hollow phrases that do only emotive work for us." Notice that to Laudan, even *unscientific* is going too far.[43]

Michael D. Gordin, professor of history at Princeton University, begins his excellent book *The Pseudoscience Wars* with a similar lament about the term: "As is surely obvious, 'pseudoscience' is a term of abuse, an epithet attached to certain points of view to discredit those ideas, complemented by 'pseudoscientist' to designate the practitioner."[44]

In contrast, the eminent philosopher and physicist Mario Bunge of McGill University in numerous books and articles uses the term *pseudoscience* without apologies and with considerable analytical force (more in the next chapter).[45] The same can be said for my philosopher/biologist colleague Massimo Pigliucci of City University of New York in his books *Nonsense on Stilts* and *Philosophy of Pseudoscience* (coedited with Maarten Boudry).[46] I think they are in line with how most scientists use it. I don't think research scientists invoke the word very often; working scientists are intensely focused on real science, with only occasional looks at the flip side of science the public finds so fascinating, but they don't agonize over it when they do. Philosopher of science Philip Kitcher also seems to be aligned with the scientific usage: "If a doctrine fails sufficiently abjectly as science, then it fails to be science. Where bad science becomes egregious enough, pseudoscience begins."[47]

Once again, in Truzzi's entry "Pseudoscience," he writes, "Because of its pejorative and sometimes pre-judgmental character, the term 'pseudoscience' is eschewed by many current analysts."[48] He would have scholars use gentler, "more neutral" terms to describe what he calls "deviant ideas" seeking scientific acceptance. The terms he proposes include *marginal science*, *unorthodox science*, *nonestablishment science*, *frontier science*, and *pro-*

toscience. He uses these terms quite often, but other than perhaps *unorthodox science*, which some scholars and writers usefully employ, the others haven't gained any general favor in the scientific community.

And as for the slam that the term is judgmental, well, scientists are in the business of making judgments. They assess the reliability and quality of scientific evidence—and assertions to scientific evidence (including their own)—all the time. They do it continually, automatically, as part of the normal processes of scientific thinking. Otherwise, they'd give equal time to every single new idea that comes along without being able to prioritize their focus on those most likely to be testable, to be productive, to be fruitful, to be most possibly correct. Some sociologists (fortunately not all) don't seem to like the fact that science makes judgments, but when the judgments are based on knowledge, experience, and well-informed analysis of the scientific evidence (or lack thereof), they are justifiable. That is just what scientists do—and should do.

Let's grant one thing upfront: Nobody calls oneself a pseudoscientist. No one undergoes years of scientific training in the sciences with the intention of becoming a pseudoscientist and doing pseudoscience. No man or woman wants to be called that or to *be* that, at least under those designations. There are scientific societies for nearly everything, but there is no National Society for the Advancement of Pseudoscience. The term is indeed a label others (usually but not always scientists) apply to designate that what they see going on is, well, pseudoscientific. Is that wrong? Is it unfair?

Other terms are frequently used. Martin Gardner was fond of referring to certain dubious work as "crank science" and "eccentric science." I think those are fine terms, but as you can see, the ever-critical Truzzi didn't approve of *crank science* either. Gardner's second big book after *Fads and Fallacies in the Name of Science* is titled simply *Science: Good, Bad, and Bogus*. That is another solution: contrasting good science with bad science and bogus science.

I like the term *fringe science* and usually think of it as being on the less egregiously bad side of pseudoscience. *Nonscience* is sometimes used, but as I've already mentioned, many if not most quite respectable human activities are not science and are therefore nonscience. Carl Sagan refers to "Night Walkers and Mystery-Mongers," his evocative title for a memorable chapter in his book *Broca's Brain*. The phrase is powerful and euphonious—I love it!—but I doubt if many practitioners of pseudoscience would want this label applied to them. It is even less complimentary than *pseudoscientists*. *Pathological science* is the term used by Irving Langmuir in a famous essay on these matters, but it hasn't gained broad currency. Robert Park uses the term *voodoo science*. I doubt practitioners and critics would much like that either.

I agree that it is usually best to just describe a person's work and how it differs from real science, without characterizing or labeling it in any way. That is actually what most scientists and science-oriented skeptics and writers usually do. They are not particularly in love with the term *pseudoscience*. They know that it has power and should be wielded with discretion. So I agree with those who say *pseudoscience* is a pejorative term. But that is not the only issue to consider. Rather than "Is *pseudoscience* pejorative?" two other questions need to be asked: Is it accurate? Is it useful, even essential?

If in any particular case, after careful investigation, a thoughtful, fair-minded, scientifically informed investigator finds it accurate to say something is pseudoscience, then it becomes more a matter of communications strategy whether to employ the word or not. And despite their reservations about its use, almost all philosophers and other scholars who deal with these issues seem nevertheless to find the word useful, perhaps even indispensable. Note that after lamenting that *pseudoscience* is a content-free epithet and is used only when scientists perceive themselves and their ideas to be threatened, historian Gordin goes on to use the term throughout his fine 2012 book (mainly about Velikovsky), including as the key word in its title, *The Pseudoscience Wars*. And he uses it again in his 2021 book *On the Fringe: Where Science Meets Pseudoscience*. And despite Truzzi's protestations about the term, the title of his fourteen-page-long (!) encyclopedia entry is simply "Pseudoscience," not "Claims and Methods Erroneously Regarded as Scientific" or something else. The word does have value. Sometimes nothing else quite works. Martin Gardner writes, with his usual clarity,

> No one can define exactly what is meant by words such as *pseudoscience*, *crank*, and *crackpot*. The reason is simple. There is no way to define anything outside of pure mathematics and logic, and even there some basic

> terms have extremely shaggy edges. It does not follow that colloquial terms assigned to portions of continua are not useful.[49]

So let's grant that all words are imprecise to some degree. Let's accept the imperfections of the word *pseudoscience*, acknowledge its power *and* its limitations—*and* its usefulness—and get on with it.

CHAPTER FIVE
THE SUBJECTS OF PSEUDOSCIENCE

When I joined the staff of *Science News* magazine in Washington early in my career as a science journalist, someone handed me a nicely printed sheet of hints on writing science. On the back of this sheet was a small-type, two-column section, "Stories That Should Be Handled with Care." It was written around 1950, or perhaps even earlier, by Watson Davis, then the director of Science Service and editor of what was then called *Science News Letter*. Davis was a pioneering science journalist. He and his predecessors and colleagues at Science Service and at several respected major newspapers and magazines had helped invent the profession of science journalism in the United States. (*Science News Letter* became *Science News* in March 1966. In 2006, Science Service became the Society for Science and the Public, now often shortened to Society for Science. In 2008, *Science News* moved from weekly to biweekly publication. Throughout 2021–2022, *Science News* and its publisher celebrated their one hundredth anniversary, a rare event in magazine publishing.[1])

The "Stories That Should Be Handled with Care" section began,

> Stories on this list should, in general, *not* be used, at least until they are thoroughly checked and investigated by several competent specialists in the subject. They are not forbidden stories, for some of the impossible things of today may become possible tomorrow, but scientific discoveries rarely come nowadays from accident or inspiration. They are usually the result of systematic research by many investigators.

The top of the list included "sweeping claims of any sort," rediscoveries of "lost arts," any "secret" scientific or technical process, and complaints by an inventor of a "conspiracy of silence."

It then listed, by category, specific topics to be handled with care. The first: "'Supernatural' Stuff." Here was included (among others) "telepathy and mind reading," "spirit manifestations of any sort," "astrologists and horoscopes," "end of the world predictions for the near future," "numerology," and "rediscoveries of lost prophetic books." Next came the "Medical" category. Included here were hypnotic "cures," any absolute cure of any disease, drugs for curing obesity and underweight, rejuvenation, electrical treatments for serious disorders, mineral waters as cures for diseases, and unauthenticated treatments of cancer, tuberculosis, colds, and diseases.

Under "Physics and Mechanics" came perpetual motion, machines that produce more energy than they use, fuel-less motors, rediscovery of such supposed lost arts as hardening of copper, "death rays," divining rods, and intuitive methods of discovering water, oil, and minerals. Under "Animal and Plant World" were listed creation of life, spontaneous generation of life, sea serpents, prehistoric and gigantic animals living today, inheritance of applied characters, and living "missing links." Concluding the list was a "Miscellaneous" section that included messages to or from Mars; skeletons or mummies of "giants" more than seven feet tall; squaring the circle; the moon's influence on weather, crops, or people; lost continents; discovery of ciphers in old books and manuscripts; and discovery of the secret of the pyramids, sphinx, or other ancient monuments.

Now, you can easily object to a list like this: "But, what about . . . ?" You can insert here you own example of some subject once thought impossible that may now not be. One of my favorite all-time books is Arthur C. Clarke's *Profiles of the Future: An Inquiry into the Limits of the Possible*. Clarke was one of my heroes, and I'm proud to own a personally inscribed copy. The introduction and first two chapters are brilliant cautionary tales about past failures of nerve and imagination by certain scientists who were trying to anticipate the future and were wrong. Some things were spectacularly wrong. (One quick example: In 1956, Britain's new Astronomer Royal, Richard van der Riet Woolley, told the press, "Space travel is utter bilge." Sputnik was launched into space the next year, and one year after that he was a leading member of the committee advising the British government on space research.) Scientists don't want to appear foolish like this. Most are loath to say that something is impossible—with the exception of when something violates the laws of physics, and even then, they might quibble. They usually prefer a word or phrase with a little more wiggle room, like *highly improbable.* (Again, note Davis's careful caveat that "some of the impossible things of today may become possible tomorrow.")

In fact, Watson Davis's list serves a valuable service. The list really does include many, if not most, of the subjects that have typically resulted in sensationalized claims that seldom stand up to scientific scrutiny. This list isn't arbitrary. It was created from top science editors' long experience with such stories and seeing how tabloids and other popular publications that care little for scientific accuracy or scientific responsibility mishandle them. These stories have endless fascination to nonscientist readers and to general writers who know nothing about science. But over and over and over again, titillating stories about such matters have turned out to be wrong. Avoiding them or handling them with great care really does keep science writers who share the values of responsible science out of morasses of mistaken science that other writers fall into so often.

Whether they've heard of this particular document or not, all good science writers and editors have some sort of such a list in their heads. It may be judgmental, yes, but just as scientists must make judgments continually about what is most likely to be good and useful and fruitful science (as I've mentioned), those who write about science for the public must do the same. As long as such lists are taken seriously and thoughtfully but not dogmatically (note Watson Davis's qualifiers in his first paragraph), they are extremely useful.

I first described this list in August 1977 at the first meeting of the Committee for the Scientific Investigation of Claims of the Paranormal (CSICOP) at the old Biltmore Hotel in New York City. As editor of *Science News*, I had been invited to give a science editor's perspective on examining paranormal and fringe-science claims. Eighty-eight stories are listed on that sheet. When I spoke, the list was already well over a quarter-century old. "Looking over that list now," I said at the time, "I can find only one ('determining or controlling of sex before birth') that could be said definitely to have moved out of that category ['stories that should be handled with care'] and into the field of legitimate science, and that only in the last few years."

Now, more than four decades have passed. The list, now more than seventy years old, still stands up well. That's amazing when you think about it. Note that it was written before the semiconductor revolution; before the structure of DNA was known; before lasers, copying machines, cell phones, the internet, and social media; and before any object or person had been rocketed into orbit or landed on the Moon, Mars, or Venus. All sciences have made enormous strides, yet I would judge that only two other items on that list have *possibly* now moved into the realm of legitimate science.

"Long-range weather forecasts in general" is perhaps one. Weather forecasting has improved dramatically since the midtwentieth century, and now three-, five-, and seven-day forecasts to some degree of accuracy are possible. Ten-day and thirty-day outlooks are made but are very general. But whether you consider any of these long-range forecasting or not is another matter. (Climate prediction, which does look at very long-term trends, is not at all the same as forecasting day-to-day weather.) "Death rays" (high-powered lasers?) is *possibly* another. Yet even here, the cautionary caveats are still worth remembering. Handle with care.

That precautionary list of stories to handle with care certainly helped shape my attitude toward such claims. Again, they aren't forbidden topics; they are just subjects where, experience shows, scientific inquiry usually leads to dead ends and where, in news organizations not well versed in scientific understanding, sensationalism often prevails over good science.

I think it may be useful to prepare our own new list of such topics and to organize them by the sciences they shadow—or at least by the sciences they share some subject content with. They are topics that people often accept uncritically and in which unscientific or pseudoscientific thinking is prevalent. One essential point I must strongly emphasize: They *can* be examined responsibly and scientifically by people who use the tools of scientific skepticism. But too often, the opposite occurs. These topics have such a strong emotional appeal that followers quickly become fans or advocates. They come to believe, often very earnestly, that the claims associated with them are, in some sense, scientifically true. At some level, their followers seem honestly to think that their pet topics are legitimate scientific advances or even whole new sciences—even if they happen to notice that most scientists seem not to share their enthusiasm.

Here, then, is my attempt to organize representative pseudosciences—or topics that typically attract pseudoscientific thinking—by the scientific fields they "shadow":

Archaeology and Earth Sciences: ancient astronauts, ancient inscriptions, the Bermuda Triangle, dinosaur and human footprints together, dinosaurs contemporary with humans, dragon hoaxes, fire-breathing dinosaurs, global flooding in human history, a hollow Earth, lost ancient technologies, lost continents (Atlantis), psychic archaeology, psychic earthquake predictions, undersea "pavements," unverified early visitations to the New World

Astronomy and Space Sciences: alien abductions, alien artifacts, alien visitations, astrology, big bang rejection, cities on the Moon, crashed saucers, an electric universe, end-of-the-world apocalypses, a face on Mars, full-Moon effects, Moon-landing denial, neoastrology ("Mars effect"), recovered saucers, rogue planets, UFOs, Velikovskyian "Worlds in Collision"

Biology and Anthropology: antievolutionism, birth-date-based biorhythms, bogus fossils (the Piltdown man), cattle mutilations, chupacabras, creationism, cryptozoology or undiscovered large animals (bigfoot, yeti, Loch Ness monster, or lake monsters, etc.), intelligent design

Cognitive Science and Neuroscience: mind-body dualism, near-death experiences, out-of-body experiences, pop psychologies about the brain, seeing Heaven, split-brain exaggerations, visiting Heaven

Medical Sciences (Pseudomedicine): acupuncture, alternative medicine, anthroposophic medicine, antiaging creams, applied kinesiology, aromatherapy, ayurvedic medicine, chelation therapy, chiropractic (other than for treating back pain), coffee enemas, colonics, complementary medicine, cleanses, crystal healing, cupping, detoxification, ear candling, electrodermal screening, energy healing, energy medicine, essential oils, fad diets, faith healing, feng shui, flower remedies, food supplements, herbal remedies, homeopathy, integrative medicine, iridology, jade eggs, magnet therapy, meridians, natural remedies, naturopathy, oil pulling, oxygen therapy, performance-enhancing bracelets, psychic surgery, quackery, quantum medicine, quantum quackery, reflexology, spontaneous human combustion, therapeutic touch, unproven medical remedies, weight-loss schemes

Physics and Chemistry: accelerators creating mini black holes, anthropic principle misinterpretations, antimatter pseudoscience, Bible codes, blood of Januarius, bomb-detector devices, cold fusion, dowsing, energy catalysis (e-cat), energy healing, faster-than-light travel, free energy, human-presence-detector rods, Kirlian photography, magnetic healing, misapplications of quantum mechanics to the macroworld, New Age physics, perpetual motion, psychic photography, quantum mysticism, relativity denial, the shroud of Turin, water with memory, weeping statues, young-Earth creationism

Psychology: aura reading, bogus self-help schemes, Dianetics, divination, eye-movement desensitization and reprocessing (EMDR), facilitated communication, fringe psychotherapies, fortune telling, ghosts, graphology, hauntings, hypnotic age regression, mass hysterias, mediums, multiple personalities, parapsychology, past lives, pop and fad psychologies, premonitions, psychic claims, psychic detectives, psychic powers (ESP, precognition, psychokinesis), psychics, rebirthing, recovered memories, reincarnation, repressed memories, remote viewing, Rorschach inkblot tests, Satanic-ritual-abuse rumors, spirits, thought-field therapy, transcendental meditation

Yes, that is quite a list! And yet it is not all-inclusive. Dozens—perhaps hundreds—of additional subjects could be listed just as well.

There is a fair amount of arbitrariness in these groupings. Many topics might just as well be classified under another scientific field, which is hardly surprising. These broad categories of science, although long accepted, are hardly self-contained. One science blends into another pretty much seamlessly. Much science today is interdisciplinary, and scientific insights from several fields are usually brought together in studying scientific ideas, so why wouldn't the same be true in the examination of pseudoscientific claims?

Let's just take one example: creationism. Creationism is usually thought of as organized political or religious opposition to biological evolution by natural selection—the fundamental insight of the life sciences. But biological evolution is central to the Earth sciences, as well. The geological ages of the planet are determined by pronounced, relatively sudden changes in the most prevalent life forms as seen in the geologic record. Geologists, paleontologists, and geochemists all helped establish the record of evolutionary change. Physics discoveries about rates of radioactive decay helped establish the ancient age of the Earth (and the cosmos). Outside of Earth history, the Sun and Solar System, the stars, the galaxies, and the universe itself have all evolved—not through biological evolution, of course, but through series of scientifically recognized sequences. (Creationists more and more dispute astronomical science, as well.)

In addition to some arbitrariness, I also readily grant something else: Almost everyone reading this list for the first time, no matter how intelligent you are and no matter how much you think of yourself as critical thinker, will find some pet subject of yours listed. You'll think it shouldn't be there. It is unfairly associated with all those other topics. You may agree that *those others* are justifiably listed, but your pet subject should not be there!

Two points to make about that. One is a concession: Some of these topics have both reasonable components and dubious components; I already mention that some parapsychological research, for example, is done by trained, conscientious experimenters—and some is not. The second is that no two people have the same pet subjects; the topics you feel shouldn't be listed here are likely to be different from those some other random person thinks have been unfairly and unjustifiably listed. You may think your pet topic has been tarred with the same brush as all the rest, but another person might instead spare the tar from her completely different subject.

There is also a broader problem with grouping typical pseudoscientific topics under specific sciences. When I first got into the business of examining pseudoscientific and paranormal claims, I came at it with a bias fairly typical of many science types: that physics and the physical sciences reign supreme. They, after all, deal with the most fundamental, basic laws of science: the "laws of physics." Aren't they the bedrock of all science? Everything else seems subordinate. I think that is true when you are dealing with scientific principles, but I am discussing our fellow human beings here. And we humans are fascinatingly complex and contradictory creatures. We all hold multiple contradictory ideas in our heads. It is a very human trait, boisterously welcomed by Walt Whitman in "Song of Myself": "Do I contradict myself? / Very well then I contradict myself, / (I am large, I contain multitudes.)"

When "flying saucers" first made their way into the public consciousness, in the summer of 1947, it was reasonable to think such sightings of something real in the skies might be some sort of secret US, Soviet, or even German (left over from the Nazis) experimental aircraft. It took a while before people started equating *flying saucer* with a supposed extraterrestrial spacecraft. In those early years, it was quite reasonable to assume that aerospace specialists and astronomers were *the* relevant experts in considering the claims. If the things were real objects flying in the skies, then aerospace engineers and astronomers were just the people to provide a nuts-and-bolts and lights-in-the-sky perspective.

The first great UFO debunker, my late colleague Philip J. Klass, was an electrical engineer and career-long writer and avionics editor for *Aviation Week and Space Technology*, the world's premier trade journal in that field. Donald H. Menzel, one of the early prominent UFO skeptics, was a noted astronomer and astrophysicist, a member of the National Academy of Sciences, and director of the Harvard College Observatory. His first book on UFOs, published in 1953 and simply titled *Flying Saucers*, emphasizes the optical-mirage aspects of the subject. But after years, and then decades, of inquiries and investigations turned up little (except in the minds of believers) to support the idea of human-made or extraterrestrial craft in the skies as the source of flying-saucer and UFO reports, attention turned more to the human dimension.[2] By the time Menzel wrote a third book on the subject, *The UFO Enigma*, he had a coauthor, Ernest H. Taves, who was both a psychoanalyst and a researcher of parapsychology and visual perception.[3] Together they were able to show, in a variety of cases from the preceding decade, the *real* cause of people "seeing" flying saucers. The human, psychological aspects of sightings were now part of the understanding.

While I was writing these paragraphs, I worried that Phil Klass, an engineer and first-class UFO investigator, might have disagreed with my point about the need for psychologists in UFO investigations, so I was delighted, upon a recent rereading of his classic book *UFOs Explained*, to find this passage near the end:

> Today . . . the UFO mystery emerges as a more complex issue that involves both physical science and human psychology. For this reason, the average physical scientist is seriously handicapped as a UFO investigator both by training and experience. . . . An experimental psychologist, experienced in the inherent limitations of human perception and recall, would place far less credence in UFO reports from pilots, police officers, and airport-tower operators than would a physicist or an astronomer, and with good reason.[4]

So Klass predated my comments by four decades, but the point bears this retelling.

In 2021–2022, there was a new frenzy over Navy pilot sightings and videotapes of unexplained aerial phenomena (UAP), the new sanitized euphemism for UFOs. The Pentagon was even pushed to form a new office to investigate the sightings. Once again, credulous UFO promoters strongly pushed the idea of tangible objects in the sky that might be dangerous or unexplained, yet in all their rhetoric and in most official government responses about this (including a May 2022 congressional hearing), very little was said about the human psychology of (mis)perception and belief. And there was little acknowledgment of UFO skeptics' various explanations of the videos.[5] Have we gone backward? Perhaps.

My general point is this: In examining pseudoscience, the human equation is powerfully important. To explain virtually any topic in our list and to understand its appeal, we need insights not just from the physical sciences but also from the behavioral sciences. Some understanding of human perception, human needs and desires, and the workings of the human mind are essential. How else to understand? Spinoza writes, "I have striven not to laugh at human actions, not to weep at them, nor to hate them, but to understand them."[6]

It took me awhile to fully realize how fundamentally crucial the knowledge and insights of our psychologist colleagues are. Theirs really is the core academic discipline in investigating pseudoscientific thinking and claims. Psychologists and cognitive scientists contribute so much. They examine not only those claims that typically fall under their domain, like psychic powers, but also all anomalous claims that tend to attract pseudoscientific thinking. Psychologists understand the nuances and pitfalls of human perception, the way our beliefs and preconceptions and expectations influence what we see, the fallibility and creative power of memory, and the methods by which we all fiercely tend to protect our core beliefs from challenges of contrary evidence.

They've achieved this understanding in modern times not just through systematic observation but also from controlled laboratory experiments. (One fascinating experiment shows how easily false memories can be implanted; the researchers actually did it.) They have shown that we find ways to discount or discredit evidence we don't like, and we rationalize doing so in ways we are not even conscious of, or if we are, these ways seem to us quite rational and proper. It is called motivated reasoning.[7] This is a natural human tendency; we all do it to some degree, but a solid understanding of psychology can make us more aware of these traits in ourselves and others. The result is a better understanding of how misperceptions arise and misinterpretations and misconceptions form. These insights make us better observers and critical thinkers.

The social sciences likewise have much to offer. They draw on much of that same understanding to study the behavior of people in groups. Sociologists help us understand the dynamics of how beliefs form, solidify, and spread. And the same with conspiracy rumors. Sociologist Jeffrey Victor argues *rumors* is a better term than *theories* because they all start with a rumor. They can't be fully understood without acknowledging how and why people in groups behave to spread them so quickly.[8] Folklorists help us understand the power of mythmaking and how societies generate narratives that reinforce our deepest beliefs, desires, and cultural values—sometimes with supporting evidence, sometimes perhaps not.

What I hope I've shown is this: The proper examination and understanding of pseudoscience is an interdisciplinary enterprise drawing on all the sciences and all the scientific, psychological, and social understanding we can muster. But it goes beyond that. Despite science's breadth, good science is not all that is required. Other academic fields are equally important. Knowledge of the history of science can help us appreciate the insights earlier inquirers achieved, the obstacles they faced, and how ideas developed. The philosophy of science makes us realize the difficulties of what it means to know something is true and to consider the complex question of what truth itself may (or may not) mean. Philosophers have explored and debated in depth virtually every issue about how science works (or sometimes doesn't), what we mean by knowledge and truth, and the differences between science and pseudoscience.

Still other fields of expertise are invaluable, as well. I mention just two: conjuring and forensics. Those who know nothing of skeptical investigation into pseudoscientific claims often belittle conjuring. That shows their ignorance. In fact, knowledge of the craft of conjuring—magic tricks, as we often sometimes demeaningly call what magicians do—is a key way to identify deception by those who claim special or paranormal powers. Magicians are *the* experts in deception. (Note: The more proper term is *conjuring*, which is the word widely used and understood in the United Kingdom; in the United States and Canada, conjurors are referred to as magicians and their art as magic. Either term works, though, and in his book on the art, James Randi uses the terms interchangeably.[9])

Magicians employ the tools of deception to make what many might consider miraculous events seem real. Of course, they do this within the context of a performance. Most make no pretense to special powers other than enormous skill and specialized knowledge of human behavior. Magicians know all the ways to make something seem paranormal (beyond science) or otherwise impossible, using their time-honored skills. They know how easily it is to fool even highly intelligent people. They do it all the time, every performance. And we know we are being fooled. Magicians with a public interest in exposing pseudoscience are therefore valued, honored, and respected colleagues of those involved in this effort. They know things that most scientists and academics, no matter how many doctoral degrees and honors they may have attained, are little aware of.

The key exemplar in this area is our esteemed colleague—and my longtime friend—James ("The Amazing") Randi. He selflessly devoted the last half-century of his life to exposing frauds and charlatans and educating scientists and the public in how best to test their claims. He gained enormous respect from scientists, academics, skeptics, and the general public for his extraordinary efforts in this area. (He was honored by the American Physical Society, and he took part in and advised published studies carried out by scientists.) Before him, the famed magician Harry Houdini did much the same thing. Houdini's book *Miracle Mongers and Their Methods*, written in the 1920s, chronicles some of his exposés of spiritualists and psychics. (Randi wrote the foreword for the 1981 reprint.[10])

Penn and Teller befuddle audiences on the Las Vegas stage and on television with their magic and yet make clear any mystical interpretations would be BS. (They even produced a debunking TV series called *Bull Shit!*) In their current TV show, *Fool Us*, Penn Jillette (the tall one) and Teller (the silent one) invite magicians from all over the world to come and do their best in an attempt to fool the hosts. Then the two discuss what they have just seen. When Penn responds to the onstage performer, he is careful to speak in a kind of magician's code so only that person knows for sure whether the hosts understand how he or she did the trick. Nevertheless, sometimes Penn and Teller perform a magic act of their own and then—reveal how it was done! This makes some fellow magicians uncomfortable, but by doing so, Penn and Teller demonstrate their commendable commitment to public education, ensuring we never consider even astonishing magic tricks evidence of the paranormal or supernatural.

So knowledge of the conjuring arts is central, not peripheral, to the examination and exposure of much pseudoscience. Many psychologists and even some other scientists and teachers involved in exposing pseudoscientific claims have some background in magic or have trained themselves in that art well enough to know what to look for. As Martin Gardner concludes his first-ever column for the *Skeptical Inquirer*, "Am I saying that all psychic researchers should be trained in magic, or seek the aid of magicians, before they test miracle workers? That is exactly what I am saying. The most eminent scientist, untrained in magic, is putty in the hands of a clever charlatan."[11]

A related field is often termed *skeptical investigation*. Skeptical investigators can come from any background, but the most successful have an intense curiosity and determination to find out the facts behind alleged mysteries. They don't presuppose an answer but use a broad range of skills to carry out field investigations and determine reliable facts about claims of alleged mysterious, miraculous, or paranormal events—everything from spontaneous human combustion to ghosts; haunted houses; poltergeists; unknown animals; and religious relics, like the shroud of Turin and weeping statues.

One outstanding practitioner of all this—perhaps the best in the world—is my colleague Joe Nickell. Joe brings an astonishing range of experience and knowledge to this quest: He has been a private detective for the Pinkerton Agency, so he can apply Sherlock Holmes–like field-investigative techniques to claims. He has done forensic investigations and worked with forensic scientists. He has been a stage magician, so he has professional experience in deception and misdirection and all the psychological techniques conjurors use. He has a PhD in English, with an emphasis on literary investigation and folklore, and is a certified document examiner, so he can bring those skills to the fore in examining the veracity of written documents. He is an accomplished author and scholar, and his awareness of the scholarly literature and knowledge of the history about myriad types of claims is broad and deep.

Whenever he can, he goes to the scene of an alleged sensational or miraculous event (recent or in the past) to see the scene for himself, interview any witnesses, and talk to other people involved. He reads contemporary news reports, examines any scholarly literature relevant to the subject, and then uses his powers of reason and deduction to put all that together to come to some assessment and judgment about the reality of the original claim. Sometimes he can come to quite clear conclusions, perhaps even debunking a claim. Other times, he raises key questions or puts the subject in a more reasonable, responsible perspective. It is quite an investigative and intellectual process. Note that he doesn't presuppose what the result of the investigation will be. It is crucial when investigating unexplained mysteries to be open-minded to the evidence and allow it to lead to any conclusion.

Many people around the world call themselves paranormal investigators (some laughably so), but only a few handfuls of them—Joe, Massimo Polidoro in Italy, Benjamin Radford, Michael "Marsh" Marshall in the United Kingdom, Susan Gerbic, Kenny Biddle, Richard Saunders in Australia, and a number of others—bring true scientific thinking to the task and are legitimate scientific investigators into pseudoscientific and paranormal claims.

Other Ways to Look at Pseudoscience Topics

I find it amusing—and instructive—to look at the topics of pseudoscience in still another way. I sometimes think of all these topics as arrayed like raisins throughout a huge piece of cake or some other three-dimensional chunk that you can cut through at various angles. If you slice in one direction you expose *hoaxes*. If you slice another direction, you see *conspiracy theories*. If you slice still another way, you see the typical topics of *denialism*. Let's take each in turn.

Hoaxes

Many pseudoscientific ideas and claims have turned out to be outright hoaxes. Here are a few examples.

Spiritualism

Spiritualism got its start in 1844 when two young girls, Katherine and Margaret Fox, in upstate New York (Hydesville, near Rochester) began talking about rappings they heard in their bedroom at night. The raps were said to be knocks from the spirits of the dead. The girls went on tour exhibiting the rappings at popular public seances. Physicist/chemist Sir William Crookes, one of several prominent scientists of the time who became enamored of such wonders, tested the girls and could find no trickery or mechanical devices for the rappings—but then he was not a magician!

Four decades later, in 1888, the two girls confessed that it was all a hoax. They had produced the rapping sounds by cracking their toe joints, a skill they had developed. But of course, the hoax revelations never caught up with the story of knocking spirits—believers either never heard of the confessions or discounted them as possibly forced lies—and spiritualism continued on its merry way under its own momentum.

The Shroud of Turin

The shroud's provenance goes back only to the fourteenth century (around 1355 AD), and the first investigation into it, in a report sent to the pope in 1389 by a bishop, Pierre d'Arcis, was revelatory—and not in a spiritual sense! The report said the cloth had been "cunningly painted, upon which by a clever sleight of hand was depicted the twofold image of one man, that is to say, the back and the front, he falsely declaring and pretending that this was the actual shroud in which our Savior Jesus Christ was enfolded in the tomb." In other words, the shroud was a fourteenth-century hoax. One might think this would be enough to condemn the shroud story to oblivion, but of course, the impulse to believe in the miraculous is exceedingly powerful. Modern forensic tests show that the image was made with tempura paint, further reinforcing that conclusion.[12]

So a shroud that almost certainly had its origin in a medieval artist's studio is still today revered by pious believers who constantly seek new rationales to claim for it some scientific validity, all the while dismissing all scientific evidence to the contrary. The best example of such scientific evidence is the radiocarbon dating of shroud cloth samples by three world-class scientific laboratories (Arizona, Oxford, and Zurich) in 1988. These accelerator mass-spectrometry tests, published in the journal *Science*, found the cloth's age to be somewhere in the range of years 1260–1390—consistent with the time of the forger's confession. "These results therefore provide conclusive evidence that the linen of the Shroud of Turin is mediaeval," conclude the three scientific groups in their joint published report.[13]

Objective, well-designed, peer-reviewed scientific tests like these have especially strong standing and credibility among scientists and among all others who appreciate the methods of science. They are a kind of gold standard. Yet believers continue to make false and special-pleading criticisms about the experiments in an attempt to persuade themselves and others that the scientists got it all wrong. They have persuaded some of the public but not most of those who are scientifically informed.

Roswell Crashed Flying Saucers

The initial impetus for the Roswell story was *not* a hoax but a misinterpretation of an actual event: the discovery of debris from an array of scientific equipment carried high by a dozen or more small balloons for a Cold War scientific experiment. The materials were borne by the winds northeast toward the New Mexico ranch, where a few days later they were found and turned in to the sheriff's office. There the misinterpretations began, some understandable, some not so excusable. But decades later,

UFO-believing writers began a series of mythmaking exaggerations. This culminated in the 1990s when some very unscrupulous promoters carried out at least three out-and-out hoaxes: the MJ-12 papers hoax (shown by Philip J. Klass definitely to be forgeries), the alien autopsy hoax, and the alien artifact hoax.[14]

The Aztec Crashed Saucer

This was the claim that a flying saucer crashed in northwestern New Mexico in 1948 near the town of Aztec, with the bodies of sixteen three-foot-high crew members, presumably from Venus. (Ironically, I live and work about midway between the sites of the supposed Roswell and Aztec saucer crashes, in Albuquerque, New Mexico, by comparison a welcome island of seeming rationality.) The Aztec story is a hoax. It was concocted by two conmen, Silas Newton and Leo GeBauer, as part of a swindle to sell devices to find oil and minerals. This has been known since at least 1952, when San Francisco journalist J. P. Cahn published a definitive and lengthy exposé in *True*. (I have a copy of it, and it is devastating.) Nevertheless, unscrupulous writers keep trying to reinvigorate the totally made-up story. And, as with Roswell, town leaders actively encourage the interest to boost tourism and visits.[15]

Here are a few other prominent hoaxes that have pseudoscientific beliefs and thinking at their core: the Cottlingly fairies hoax, the Loch Ness monster (the 1934 photo hoax), the Piltdown man, King Tut's curse, the Amityville horror, psychic surgery, crop circles, the 2012 Nibiru doomsday hoax.

Hoaxes have always had a substantial role in the spread of pseudoscientific thinking. Learning about them makes us appreciate how far some people will go to deceive the public. Sometimes they may do it just for fun, to see what they can get away with and how far the story will spread. Sometimes the motives are more sinister. Their motivation doesn't matter. Large segments of the public are deceived. The good news is that the exposure of hoaxes can awaken us to the prevalence of intentional deception and the ease with which we can be deceived. That's an important lesson to us all in critical thinking, so we can better be on guard for the next one.

Conspiracy Theories

Conspiratorial thinking has always been an undercurrent of modern culture, but in the past decade or so, it seems almost to have gone mainstream. The internet makes it exceedingly easy to spread conspiratorial memes, and the general atmosphere of mistrust that permeates our times provides fertile ground for ideas of conspiracies to take root. There are, of course, real conspiracies and always have been; critical thinking is what's needed to help us know which are real and which are products of the imagination. Here is a partial list of conspiracy theories that, from a scientifically informed viewpoint, fall into the latter category.

QAnon

QAnon, the ultimate conspiracy theory, the mothership, the "choose your own adventure" saucerful of codes and secrets. QAnon first appeared in October 2017 on the message board 4chan, but it burst into public consciousness during the global pandemic, which reinforced followers' beliefs in an oncoming but nebulously defined "storm." QAnon found allies among antivaxxers, antimaskers, and antilockdown protestors. A core belief among many was that hundreds of thousands of children were being kept locked up in cages and tortured by armies of pedophiles. The Pizzagate conspiracy was an early indicator.

And then President Trump started proclaiming, despite all evidence to the contrary, that he, Trump, had actually won reelection on November 4, 2020, and that his election had been stolen from him (Joe Biden won by seven million votes). This proved irresistible fodder for many conspiracy believers. A number of prominent QAnon figures were among those who broke into the US Capitol building on January 6, 2021, to try to stop Congress's certification of Biden's election as president. Some later recanted their views; many were convicted or pled guilty. It was the largest prosecution in the history of the US Department of Justice.

By early June 2022, more than eight hundred people had been charged with crimes, and more than three hundred people had pled guilty to such crimes as conspiracy and assault. Nearly two hundred people have been sentenced so far.[16] Whether most participants were part of QAnon or not (many were part of the extremist right-wing antigovernment Proud Boys and the Oathkeepers), almost all believed to some degree the conspiracy theory that the election had been stolen from Trump.[17]

The COVID-19 Vaccine

During the pandemic, dozens of conspiracy theories were concocted about how the COVID-19 vaccines were quickly created and distributed. These spread like wildfire on social media and on radio talk shows hosted by unscrupulous conspiracy theorists and swirled around the world in 2020–2021. Like the damage done by the earlier conspiracy theory that vaccines cause autism, the COVID-19 vaccine conspiracy theories resulted in major resistance against the life-saving vaccines and inhibited the administration of the vaccines to millions of people. This campaign kept case rates and deaths far higher than they otherwise should have been.

A December 2021 Annenberg Science Knowledge survey found that 31 percent of conspiracy theorists believed that the Chinese government created the coronavirus as a biological weapon (there is a similar conspiracy theory that the AIDS virus was deliberately created), 21 percent believed that Bill Gates supported a vaccine containing microchips that can track a vaccinated person, and 15 percent thought that the Moderna and Pfizer vaccines contained fetal tissue.[18]

The 9/11 Truth Movement

The 9/11 Truth Movement is the idea that the collapse of the World Trade Center towers on September 11, 2001, was caused by planted explosives rather than by the high-speed impact of hijacked, fully fueled jet airliners. It also includes the belief that the 9/11 attacks were carried out by the US government, Israel, or anyone but the Islamic extremist terrorists who actually did them.

World Government

There are many conspiracy theories about world governance, such as the Trilateral Commission, the Illuminati, and the Protocols of the Elders of Zion. These encompass such beliefs as the Apollo moon landings were staged and government coverups of UFOs and aliens, including the belief that alien spacecraft and alien bodies have been recovered and whisked off to secret military bases. Here mythmaking and conspiratorial thinking intersect in a virulent strand of antigovernment suspicion and general antirational weirdness. Given the recent disturbing trends of distrust of government and science, these ideas are if anything stronger today than they were two decades ago.

Which are the most popular conspiracy theories? All polls are deficient in some way, but they can give us a rough idea. In October 2021, the *Washington Post* ran an interactive article asking a series of questions to gauge beliefs in certain conspiracy theories. Here are some of the levels of belief it found:

GMOs are secretly dangerous (45 percent).
JFK was killed by a conspiracy plot (44 percent).
There is a deep state embedded in the government (43 percent).
One secret group controls the world (35 percent).
COVID-19 is a man-made bioweapon (31 percent).
Democrats stole the election (27 percent).
Trump faked having COVID-19 (26 percent).
Obama faked his birth certificate (20 percent).
School shootings were faked by the government (17 percent).[19]

Again, no single poll should be given too much credence. And *belief* is not an entirely satisfactory word. How strong is the belief in each case? This poll doesn't tell us.

Not all the trends show increasing belief in conspiracies. One that has fewer adherents than before is that NASA is covering up discoveries of cities on the Moon or alien-built stone faces on Mars. That once-persistent claim doesn't seem to have much support anymore. This is for good reason: The Moon is visible to all of us, and the Mars face has been repeatedly imaged by Mars-orbiting NASA spacecraft at higher and higher resolutions. The photos show persuasively that what originally looked like a crude face due to the play of light and shadow is really a natural formation. Under higher resolution and different light angles, the facial features disappear.

Denialism

Denialism is another mode of thinking disturbingly prominent in society today, and it is a form of pseudoscience.[20] Cut into the block of pseudoscientific topics, and you can expose an array of denialisms. All have an agenda, and to accomplish it, the adherents denigrate, undermine, and deny the best scientific evidence. Denialisms all involve the denial or rejection of some aspect of science in one way or another. Adherents reject either unwelcome conclusions of scientists or the methods of science that lead to them or both. When that fails, they attack the scientists themselves, sometimes in very vicious, personal ways. Denialism not only spreads a range of modern pseudoscientific and antiscientific ideas but also, more troubling, solidifies them into something highly impervious to contrary evidence. Denialism is like concrete. Take a mixture of belief, ideology, and scientific distortions and stir, and it quickly hardens, in this case, into something resistant to all reason and rationality.

Here are eight especially strong and even dangerous examples of denialism:

1. vaccines denial
2. evolution denial
3. climate change denial
4. AIDS/HIV denial
5. Moon-landing denial
6. ancient-Earth denial
7. big bang denial
8. Holocaust denial

Denialists hate being called denialists. They may deny the science of the subject they have issues with, but they don't want to be associated with the earlier usage of the term, the last topic on this list, Holocaust denial. I understand that. So we have another semantic difficulty. Once again, though, the issue is not whether they dislike the term but whether it is accurate. Clear thinking requires we give priority to accuracy of language, not to feel-good euphemisms or weaselly political correctness.

Another thing I'm not willing to grant, nor should any of us: Climate change denialists have fancied calling themselves skeptics. They do so all the time, but that is an abuse of an honorable term. They are expressing extreme, not conditional, skepticism of comprehensively supported, well-accepted science. By that, I mean the scientific conclusions that the Earth system is heating up and climate change is underway are based on multiple, independent sets of extensive global data and tens of thousands of published studies and analyses from many different, mutually reinforcing scientific disciplines. So what the deniers are doing is not skepticism; it is denialism. It is not substantively different from what evolution deniers (creationists) and vaccine deniers and AIDS/HIV deniers do.

Real skeptics are always willing to revise their ideas in light of new scientific evidence, but that evidence is the key. It is the scientific evidence that guides the opinion, not the opinion that decides which scientific results to accept. Some suggest a better term for climate denialists who profess (wrongly, as I have argued) that they are skeptics is *pseudoskeptics*. I think that is a reasonable description.

The Name Game in Pseudoscience

On the matter of semantics, I alert you to a related problem: The topics of pseudoscience keep changing names. Names and terms change, but the same concepts emerge in new guises. It is as though once claims have been repeatedly discredited, someone gets the bright idea to repackage them in different wrappings. The contents are the same. What was once called clairvoyance is often today called remote viewing. Parapsychology, a name in use for many decades, is now frequently called anomalous cognition. Cold fusion has morphed into low-energy nuclear reactions.

Immanuel Velikovsky's discredited "worlds in collision" ideas from more than a half-century ago are seldom mentioned anymore, but some old supporters have joined new enthusiasts with some very similar unsupported ideas about the cosmos, which they now call the electric universe. Creationists tried to popularize the term *scientific creationism*, without much success, and then updated their main new thrust two decades ago to the term *intelligent design* (without saying who the "designer" is). Creationists are especially adept at changing names and strategies in an attempt to sound respectable (and not overtly religious, which would bring them into conflict with the Constitution), but science-minded people—and most legal specialists—are not fooled in the least.

The same shape-shifting is done by practitioners and promoters of medical pseudoscience. We used to call it quackery or snake oil. Then it became unscientific medicine and then alternative medicine and now is often referred to as integrative medicine. It almost does sound respectable in those latter guises, and some medical schools have gone with the fad and adopted some of that terminology. But with only few exceptions, what is at their core are a host of untested medical remedies—or, even worse, remedies and practices that have been tested and failed. If it were otherwise, then they would be quickly adopted and become part of real medicine, and there would be no need for *alternative* terminologies.

The topics of pseudoscience are legion. In this chapter, I provide a brief overview of some of the most prevalent and some ideas about how to organize them by the sciences they shadow. Some can also be seen as hoaxes, conspiracy theories, or denialisms. Most retain popularity (despite remaining unproven) for long periods of time; a half-century or more is typical. But some change their names or outer guises regularly to seem new and different.

I caution again: All these topics *can* be examined responsibly using the tools of scientific and critical thinking. My colleagues who investigate and study them from a scientific perspective do that all the time. But experience shows that these topics tend to invoke enthusiasms and yearnings well out of proportion to their empirical validity (or lack thereof). As a result, they should raise—and *do* raise—cautionary red flags for everyone concerned about what is real. It is precisely because they have such ready appeal and such emotional power that we must guard against our own and others' wishful thinking—and still always consider them "stories that should be handled with care."

Yet this chapter is mainly a look at the skeletal structure of pseudoscience, essentially its gross anatomy. For what really makes something a pseudoscience, we must look deeper—into its functioning and dynamics, its working physiology. We need to look at what makes for pseudoscientific thinking and how that thinking contrasts with scientific thinking. That I explore in the next chapter.

CHAPTER SIX
PSEUDOSCIENCE IN MEDICINE, OR SCAM

A few weeks back, my wife and I were staying at the home of dear friends in the western suburbs of Chicago, our first visit since before the pandemic. One night, they invited a third couple to join us for dinner. Over predinner drinks, the other couple proved interesting, charming, intelligent. The husband was European born, sophisticated. When we sat down for dinner, however, the first words out of his mouth surprised me: "I had an acupuncture treatment on my back yesterday. I am so much better. I am a big believer in acupuncture." Nothing we'd said had prompted the proffered statement. "Oh boy," I thought but said nothing. Many people are enthusiasts about acupuncture.

He went on about how he goes to his acupuncturist regularly for severe chronic back pain, and the treatment always makes him feel better for at least twenty-four hours. He continued in this vein until our host, my close friend, got me in trouble. He pointed out that I was a skeptic about such matters and undoubtedly had a different view about acupuncture.

We had really just sat down at the table. I wanted to enjoy a pleasant evening. I didn't want to talk about the need to be skeptical about claims of alternative medicine. I wasn't ready to talk about what I do, what skepticism is, why acupuncture has no science behind it (other than a placebo effect), and why acupuncture patients often nevertheless feel (temporary) pain relief.

I fumbled for some words. "I'm glad you get some pain relief from your acupuncture," I said, trying to be conciliatory, "but to the degree your acupuncture is based on the idea of meridians"—and all acupuncture is—"it has no basis in science or fact." Meridians don't really exist, except in the folklore of traditional Chinese medicine. I spoke about how pain is very subjective and how undergoing a treatment meant to make you feel better can actually make you feel better for a while, but it doesn't treat the pain or its cause, just your subjective experience of the pain. Acupuncture can't really be used to treat any real organic disease. It wasn't really a good start to the dinner conversation.

"Are you saying it is psychosomatic?" he asked.

"No," I assured him. "Your pain, I am sure, is very real." But I told him that inserting needles at so-called meridian points is not what made him temporarily feel better. They could have been inserted anywhere or not at all. It is the whole experience. Some critics have talked about acupuncture as theater, a dramatic performance. And if someone is going to all that trouble to perform specifically for your benefit, then you more than likely are going to feel better for a time. Note that he acknowledged that the pain usually returned within twenty-four hours.

His wife interjected and to the relief of us all quickly changed the subject. We went on to have a delightful evening, but I wish I had handled the matter better. It is not very productive to tell someone one of their deep-seated beliefs is wrong. That doesn't usually persuade. At their departure, I apologized for sounding so bluntly critical about his acupuncture. He was very kind: "Oh, I don't take that personally."

Like pseudoscience, pseudomedicine (pseudoscience in health care) is all around us. Much of it is often called alternative medicine, a moniker that, as I show in this chapter, is both misleading and meaningless. Medical pseudoscience is popular and pervasive, even more so than most of the other pseudosciences listed in chapter 4. Few of its advocates and users have any idea that they are using treatments that have little or no scientific legitimacy. Or if they have such an inkling, then they do not care. They have already decided that the treatments—pills, liquids, concoctions, devices, whatever it is they are using—"work for me."

The conversation needn't have been about acupuncture. Acupuncture, popular though it may be, is relatively harmless in comparison with many other pseudomedical treatments. It was just as likely the dinner guest would be an enthusiastic and happy user of homeopathy. Or naturopathy. Or chiropractic. Or therapeutic touch. Or detoxifying. Or coffee enemas. (Now that would have made for fun dinnertime talk!) The alleged remedies and treatments of so-called alternative medicine—much of it pseudomedicine—are so multivarious and so prevalent that they have taken a commonplace position in the pharmaceutical menageries, seemingly not very different from the prescription drugs proved by controlled experiments and certified by government agencies to be safe and efficacious. Users see little difference between the two, and the alternative medicine industry exploits and profits mightily from that confusion.

Alternative medicine is not just one thing. I very much like a description by one of my favorite persons, Harriet Hall, a retired family physician. She had a long career in the US Air Force (where, as a flight surgeon, she not only treated patients but also flew big planes—really big planes—and afterward even wrote a book titled *Women Aren't Supposed to Fly*).[1] In retirement she took up researching and writing insightfully about medical pseudoscience and the shortcomings of alternative medicine. She is known as the SkepDoc for her appropriately skeptical attitude toward such things, and she writes clearly and directly:

> Alternative medicine embraces many things: treatments that have never been tested or have not been adequately tested; treatments that have been tested and shown *not* to work; treatments that are based on nonexistent phenomena such as human energy fields and acupuncture points; treatments such as homeopathy that if true would violate established scientific knowledge; and treatments that have been proved to work but that mainstream doctors have good reasons not to recommend.[2]

Note how this little evaluation is nicely nuanced, one indication that her appraisal is fair and reasoned. It also is a recognition of how astonishingly diverse the treatments and claims of alternative medicine are and how they so often fall short of proponents' promises. Here are a few examples Hall provides to show what she means:

> Tai chi has been proven to reduce falls in the elderly, but other more conventional treatments might be more suitable, cheaper, or more available. Kava has been proven to reduce anxiety, but it appears to be toxic to the liver and has been banned in many countries. Garlic has been proven to reduce blood pressure and cholesterol, but the effect is too small to be clinically significant, and we have drugs that are more effective

> and safer. . . . Chewing willow bark relieves headaches, but aspirin works better and is readily available, well-regulated, and inexpensive.[3]

Hall's short take on alternative medicine also demonstrates something else: Yes, it is important to acknowledge nuances and the great variety of claimed treatments and approaches of alternative medicine; it is also important to state clearly and objectively what they really are and what is wrong with them. That is an essential part of clear thinking and critical appraisal.

One more observation from Hall is worth sharing. It's about herbal medicine, which can seem highly appealing, but it could apply equally to many other alternative remedies: "Standardization is lacking in the herbal medicine marketplace, and many products have been found to be contaminated with everything from prescription drugs to carcinogens to insect parts and floor sweepings."[4] Yow! Scientists have made similar objections to other claimed remedies, as well. They aren't regulated, so you don't really know what is in them.

When it comes down to it, many critical observers agree with a generalization from Paul Offit, who, in addition to being a noted physician, professor of medicine (University of Pennsylvania School of Medicine), and vaccine expert (he was part of the team that invented the vaccine against the rotavirus), has written books about vaccines, antibiotics, alternative medicine, dietary supplements, megavitamins, faith healing, scientific discoveries gone awry, and the bad advice celebrities and activists give on health matters: "The truth is there's no such thing as conventional or alternative or complementary or integrative or holistic medicine. There's only medicine that works and medicine that doesn't. And the best way to sort it out is by carefully evaluating scientific studies—not by visiting Internet chat rooms, reading magazine articles, or talking to friends."[5]

SCAM's Basics and Ernst's Insights

Edzard Ernst is the world's most prolific researcher in testing claims of alternative medicine. He has concluded that the term *alternative medicine* just no longer works:

> We used to call it "alternative medicine," the name most people still know best. But lately I have started employing a different term for it; I now tend to call it *so-called alternative medicine* or *SCAM* for short.

> Why?
>
> Mainly because, whatever it is, it clearly is not an alternative:
>
> - If therapy does *not* work, it cannot possibly be a reasonable alternative to an effective treatment.
> - And if it *does* work, it simply is part of medicine.
>
> After having been involved in this subject for over 25 years, I feel that "SCAM" is preferable to the many vague and imprecise terms that have been used previously.[6]

He's not the only physician or medical researcher who now uses the term, but he has used it in the title of two books already![7]

In chapter 3, I talk about how pseudosciences constantly change their names and public identities (while basically remaining the same) and how medical pseudoscience has had the most monikers over the decades. (In its case, some of the terms are descriptions applied to them by mainstream medical scientists, but most—like *alternative medicine*—are names adopted by its own practitioners and the media to sound more consumer friendly.)

Here is Ernst's list of those earlier "vague and imprecise" terms, in reverse alphabetical order: *unproven medicine*, *unorthodox medicine*, *unconventional medicine*, *natural medicine*, *integrative medicine*, *holistic medicine*, *fringe medicine*, *disproven medicine*, *complementary and alternative medicine* (*CAM*), *complementary medicine*, and *alternative medicine*.

Ernst defines SCAM as an "umbrella term for many therapeutic and few diagnostic modalities that are not generally accepted as useful by conventional healthcare professionals while being promoted as helpful by practitioners operating outside the mainstream of medicine."[8] Note how that definition echoes the various definitions of *pseudoscience* I give in chapter 3, especially the two contrasting aspects: "not generally accepted" by scientists (for lack of evidence) but nevertheless promoted and advocated as real by proponents. That is why I have little trouble labeling a good part of so-called alternative medicine "pseudomedicine" or "medical pseudoscience."

Ernst wrote a book evaluating 150 highly diverse modalities in SCAM but says there are "well over 400 in total."[9] He has kindly given me approval to quote extensively from his works, and I want you to hear from the world's leading expert. Here is his much briefer list of the "most popular" SCAM therapies:

- acupuncture
- anthroposophic medicine
- aromatherapy
- chiropractic
- crystal healing
- dietary supplements
- energy or paranormal healing
- herbal medicine
- homeopathy
- mind-body therapies
- naturopathy
- osteopathy
- reflexology
- reiki
- therapeutic touch
- traditional Chinese medicine[10]

He notes that these therapies have become more prominent in recent decades because large proportions of the population use them, ranging from 25 percent in the United Kingdom to 70 percent in Germany. He is especially concerned about their appeal to cancer patients. Cancer patients are normally desperate for help and vulnerable. They can hardy ignore the "relentless promotion of SCAM."[11] You can understand their appeal: They promise a cure, whereas real medicine cannot always do so. Patients want a treatment free from risks and mistakenly think SCAM treatments offer that. They fear adverse effects of conventional treatment (very understandable), and so on.

My family was a firsthand witness to the "relentless promotion of SCAM" to cancer patients. When our daughter, Michele Baldwin, became mortally ill with late-stage cervical cancer and decided (after two rounds of surgery and chemo) to devote the last six months of her life to a first-of-its-kind expedition in India to call attention to prevention methods (pap tests and the then-new HPV vaccine), a national online news service one day prominently ran an article about her. Within minutes, her expedition website was inundated with suggestions for a variety of SCAM treatments. We were amazed how quickly they came in. (Are they automated?) "They came out of the woodwork," my wife, Ruth, recalls today. Michele was not pleased, knowing they were useless and a distraction, and neither were we.[12]

Edzard Ernst is a remarkable person. Early in his career, he practiced homeopathy and thought there might be something to alternative medicine, but he changed his mind as better evidence arrived. Armed with both a medical degree and a PhD and already a professor of medicine, Ernst was invited to become an endowed professor of complementary medicine at Exeter University in the United Kingdom, charged with carrying out a program of rigorous research to test and "reveal the true value" of alternative medical treatments.[13] The man who had endowed the chair felt he and his wife had been helped by such treatments and assumed rigorous studies would show positive effects. Ernst, for his part, said he would be skeptical.

At first, he kept his head down, formed a team, did the needed research, and published their results almost entirely in medical journals. Dozens and dozens, then hundreds of scientific studies were published. It is a remarkable body of research. Eventually he came to realize he would need to "stick my head out" and stand up publicly for his research team's findings.[14] That is because their research, as well as that by other scientists around the world, clearly showed that, in his words, "few forms of SCAM seemed to work, many seemed not to do more good than harm, and most SCAMS were so under-researched that it was impossible to tell."[15]

They eventually published a *Desktop Guide to Complementary and Alternative Medicine*.[16] By then they had evaluated seven hundred treatment-and-condition pairs. They found that the amount that was supported by sound evidence was only 7.4 percent! More than 92 percent of the treatments had no sound evidence in support. And even these numbers overstated SCAMs' validity. That is because the studies had focused on treatments for which there was at least some evidence. "Had SCAMs been included indiscriminately, the resulting percentages would have been much smaller," Ernst writes.[17]

Surely these kinds of results, surprising to many and important because the public often uses such treatments without even consulting a real physician, should be widely known. Realizing that, Ernst began writing in daily newspapers, lecturing to the public, and in general reaching out to spread the message in any way possible.

By the end of the 1990s, several independent analyses had shown that Ernst's team had become the most productive research unit in SCAMs worldwide. More scientists asked to join the team, and many of them went back to their home countries to occupy key positions and do further research. "Our concept of critical evaluation thus spread around the world," Ernst says. Around ninety researchers have worked with his team over the past twenty-five years—many, he says, are "gifted scientists"—and he says he is deeply grateful to all of them.[18]

One of the persons who had been peripherally responsible for Ernst's position at Exeter was Prince Charles, a longtime (some might say notorious) advocate for alternative medicine. The ever-irreverent Christopher Hitchens once wrote of Charles, "We have known for a long time that Prince Charles's empty sails are so rigged as to be swelled by any passing waft or breeze of crankiness and cant."[19]

Charles soon got annoyed at how the team's studies were instead showing that most alternative remedies didn't do what they claimed. He wanted research, but only the kind that had positive results, not really the scientific method. He and his supporters managed to get Ernst's research funds withdrawn, forcing Ernst into early retirement.[20] "I prefer rigorous science to wishful thinking, a stance that many SCAM proponents find hard to accept," says Ernst.[21]

Ernst retired from his regular Exeter post in 2012 and began writing a blog and books.[22] He continued to get flack for his work, but he no longer had to fear getting officially reprimanded by his peers for not being "politically correct" (his words). In 2015, he was honored with the John Maddox Prize for standing up for good science. The citation honored him for "his long commitment to applying scientific methodologies in research into complementary and alternative medicine" and for continuing his work "despite personal attacks and attempts to undermine his research unit and end his employment."[23]

Ernst notes that some of his critics claim he never did anything but debunk SCAM. That is not true. In 2008, he published a summary of all the treatments that, according to his team's analyses, were based on sound evidence. Here, for instance, are a few treatments he says "might do more good than harm" for specific ailments: acupuncture (for nausea/vomiting), gingko biloba (for Alzheimer's and arterial disease), hypnosis (for labor pain), massage (for anxiety), melatonin (for insomnia), music therapy (for anxiety), relaxation (for anxiety), and Saint-John's-wort (for depression).[24]

Doing more good than harm is key. For most other SCAM treatments, Ernst says,

> the benefits are usually uncertain, small or even non-existent. . . . This fact is crucially important:
>
> - If there are no benefits, the balance cannot be positive, even if the risks happen to be small.
> - If the benefits are small, even relatively minor risks would produce a negative risk/benefit balance.
>
> . . . Few SCAMs will ever be associated with a positive risk/benefit balance.[25]

Alternative medicine, pseudomedicine, medical pseudoscience, SCAM—whatever we call it—shares a variety of characteristics with all pseudoscience. Among them are the assumptions, excuses, and rationalizations offered by their proponents. Many medical commentators have written about this, but here are a few offered by Ernst:

- SCAM is helpful.
- SCAM is natural and therefore safe.
- SCAM defies scientific investigation.
- The "establishment" wants to suppress SCAM.
- SCAM is holistic.[26]

That the treatments are helpful is widely believed (by believers anyway), but unless they do more good than harm (see previous), they don't help. Cancer patients are especially susceptible to harm from avoiding real medicine while seeking unproven remedies. Apple cofounder Steve Jobs avoided real treatments for his pancreatic cancer in favor of various alternative remedies. By the time he realized his mistake, it was too late. He died just months later.

The assertion that SCAM is natural is a diversion. "Natural" is an appealing thought: Water and air are natural, and we can't survive without them. Arsenic also is natural, but too much of it is deadly. What is natural about sticking needles into the skin of a patient (acupuncture)? Herbal supplements may start out natural, but they end up being highly processed. The naturalistic fallacy is one of the most common errors advocates of alternative remedies make. But this appeal to naturalism is a great marketing strategy, so almost all remedies use it in some form. "The disappointing truth is that SCAM has few qualities that would truly render it natural," says Ernst.[27]

The holistic claim is especially misleading. Holism—dealing with the whole patient, not just symptoms—has always been a central element of good health care. SCAM practitioners may claim to be holistic, but they are not well trained to deal with complex conditions (like cancer), they regularly overestimate their abilities to deal with serious conditions, and they rarely refer their patients to other professionals and specialists. "Consequently," says Ernst, "SCAM is usually less holistic than the practice of modern healthcare. And less holistic can also mean more dangerous."[28]

That SCAM defies scientific investigation is also a sham claim. Just think of all the ways scientists have learned formerly unfathomable mysteries of the universe; about ourselves and our bodies, including the structure of DNA; the way proteins fold; and the deciphering the human genome (and now the genomes of many of our fellow creatures), not to mention the frontier scientific discoveries I describe in the opening of this book—including the first detection of gravitational waves and the first image of a black hole and then the first image of the black hole at the dust-obscured center of our own galaxy. These are extraordinarily difficult feats. Scientists are exceedingly clever and resourceful in finding new ways to tease out the truth about things. And of course, thanks to medical researchers like Ernst, most alternative remedies now have been subjected to detailed scientific investigations—and most have failed.

There are now thousands of clinical trials of SCAM, with a few even demonstrating more good than harm. Ernst notes that SCAM proponents do enthusiastically accept scientific investigations of their remedies—"as long as the results are positive."[29]

The claim that the establishment wants to suppress SCAM is also without evidence. Pharmaceutical companies are constantly searching for any new remedies that work. So are medical researchers and their universities, hospitals, and laboratories. "Whenever I asked a believer in the conspiracy theory of the suppression of SCAM to show me any evidence of his assumption, I ended up empty-handed," says Ernst. "There simply is no such evidence."[30]

Celebrities Hawk SCAM

Prince Charles is a case study for how certain celebrities and public figures have ardently advocated for alternative medicine. They have played an outsized role in its public popularity. (Prince Charles ascended to the throne of England as King Charles III upon the death of his mother, Queen Elizabeth II, on September 8, 2022, but I retain the "Prince" title here because all the examples are from his decades as prince.) From an early age, Charles seemed suspicious of science's materialism and was attracted to mysticism. He learned about the "collective unconscious" from a devotee of Carl Jung who was later shown to be a noted fantasist and liar.

Undoubtedly because of his fame and position in the royal family, the British Medical Association elected him as its president when he was only thirty-four. Asked to address that body, he turned his talk into a full-frontal attack on conventional medicine, accusing it of unfairly rejecting anything unorthodox or conventional and failing to address the whole patient. The physicians were aghast. He later did much the same to the World Health Organization (WHO).

Charles has never liked science's commitment to objectivity, preferring instead to rely on intuition and beliefs. Over the decades, Charles repeatedly interfered in politics, lobbying for Britain's National Health Service (NHS) to fund alternative medicine. His letters and memos about this were revealed after a protracted legal battle. Ernst says they are filled with misleading statements and misrepresentations. One is the assumption that it would reduce costs. Ernst notes wryly that there cannot be *cost-effectiveness* without *effectiveness*.

In 2008, Charles even became an alternative medicine entrepreneur himself, launching his Duchy Originals Detox Tincture. A mixture of globe artichoke and dandelion extracts, it promised to remove toxins and aid digestion. It had been tested only for safety and quality, but the company claimed it had been tested for efficacy. It was quickly censured by the UK Advertising Standards Authority for violating advertising rules.

We know all these things and much, much more because in 2021 Ernst published an entire book about Charles's "alternative" enthusiasms and misleading pronouncements, *Charles, the Alternative Prince: An Unauthorized Biography.*[31] In a long, substantive review, SkepDoc physician Harriet Hall recommends the book wholeheartedly and says it "tells the whole story of Prince Charles's ignorance and folly matter-of-factly, relying mostly on Charles's own words."[32] She said it should prove of interest to anyone who wants to learn more about science, skepticism, and critical thinking.

And in case you are wondering whether Ernst hates Prince Charles and wrote the book to get revenge, Hall says, "I don't think so. There is no hint of personal animosity in the book." She adds that Ernst is not prejudiced against alternative medicine per se; "he only asks that all claims for efficacy be based on credible evidence, regardless of whether they are alternative or mainstream."[33]

Celebrity culture in the United States has long bolstered alternative medicine. Popular and photogenic television and movie stars—understandably appealing but hardly known for their scientific understanding—have been at the root of all sorts of enthusiasms, fads, and misunderstanding. It can be difficult to counter their sincere and usually well-meaning advocacy with evidence, facts, and reason. This is a difficult communications situation for physicians and scientists. They are at a disadvantage when popular celebrities and public figures take it upon themselves to advocate what is clearly nonsense to the scientific mind.

Paul Offit, the noted physician, medical researcher, and vaccine advocate I quoted earlier, became a well-known public figure himself during the pandemic of 2020–2022, appearing regularly on national television on behalf of the vaccines against COVID-19. He wasn't as ubiquitous as Dr. Anthony Fauci, but he was nevertheless a frequent (and in my view effective) guest on many national television news shows. He was congenial. He spoke clearly and to the point, without jargon. But Offit wasn't always comfortable on television, especially when famous celebrities were involved.

Offit tells the story of his decision to appear on Oprah Winfrey's highly popular national television show *Oprah* in 2007 with Jenny McCarthy, the TV actress who had become convinced of the mistaken idea that vaccines cause autism. He thought it would be a good chance to educate millions about the value of vaccines and how the measles vaccine didn't cause autism. It didn't turn out that way, not with those two appealing personalities on the TV stage, one already world famous, one about to become so. He thought this would be a "chance to calm the waters," but he found himself up against the power of the personal narrative.[34]

The show wasn't about calming the waters. He relates the story in his book *Bad Advice: Or Why Celebrities, Politicians, and Activists Aren't Your Best Source of Health Information*:

> Oprah was there to tell a story. And her story had three roles: the hero, the victim, and the villain. Jenny was the hero. Her son was the victim. This left only one role for me. I would be the guy telling Jenny that she was wrong and that, by extension, Oprah was wrong to have had her on the show. Jenny knew the cause of her son's autism (vaccines). I didn't. Jenny had cures for autism (megavitamins, hyperbaric oxygen chambers, and mercury-binding medicines). Also, not to be completely politically incorrect, but as a male scientist, you can't go on a television show in front of a studio audience consisting entirely of women and tell two women that they're wrong. It just doesn't work.[35]

McCarthy told the audience her story about her son's autism, his MMR shot, her forebodings that she had a "bad feeling" about it. She didn't hold back on the emotion. "Fighting back tears," Offit recalls, "McCarthy explained how her son had been fine one minute and then, because of that vaccine, had been condemned to a lifelong struggle with autism."[36] It was impossible to compete with that. Offit says that later in the program, Oprah read a dry, prepared statement from the CDC stating that McCarthy's concerns weren't supported by the evidence. But who is the audience going to believe, an anguished and appealing mother or some carefully worded missive from some distant and anonymous scientists representing some monolithic institution? "The CDC didn't have a chance," says Offit. "McCarthy's heartfelt confession won the day. Jenny

McCarthy's appearance on Oprah launched her career as a powerful force against vaccines. And we have Oprah Winfrey to thank."[37]

I am a great admirer of Oprah Winfrey. She is a humane, compassionate, powerful voice, especially for women and for the downtrodden and neglected. She has done a lot of good for the world.[38] But her frequent advocacy of bad medical science instead of responsible medical science has caused a lot of mischief, misunderstandings, and worse. She also launched Dr. Mehmet Oz's career. Oz, although a physician himself (turned elected politician), has always had a passion for fringe medical ideas. Thanks to Winfrey, he soon got his own show. He used it to promote alternative therapies ranging from naturopathy, homeopathy, acupuncture, therapeutic touch, faith healing, and chiropractic manipulations to communicating with the dead. Offit says, in a sense, *The Dr. Oz Show* is a "voyage back through the history of medicine, starting with our most primitive concept of what caused disease: supernatural forces."[39] Oz later embraced the occult, including so-called psychics like John Edward.[40]

As Offit writes elsewhere in his book, doctors like Mehmet Oz and Deepak Chopra are "so seductive" because they "represent themselves as all-knowing gurus." They express themselves with certainty. Certainty is what people want. They "know the truth, and the truth is immutable, fixed." It doesn't matter that that's not how the world—or medical science—works; "when our health is at stake, it's hard to accept that our knowledge is incomplete."[41]

The late Martin Gardner, one of our great national treasures and a lifelong perceptive critic of pseudoscience, writes about Oprah in one of his last *Skeptical Inquirer* columns, "Oprah Winfrey is an attractive, intelligent woman with a heart of gold, but who has only a pale understanding of modern science. On her daily television show [which ended in 2011 after twenty-five seasons], she promotes, as frequent guests, men and women who preach views and opinions that are medically worthless and in a few cases can lead to death."[42]

Among many other things, Gardner especially notes Oprah's promotion of Jenny McCarthy and her pseudomedical causes. McCarthy vigorously promotes the idea that autism is caused by the measles vaccine. She claims that her autistic son, Eric, has been helped by chelation therapy. This therapy blames autism on mercury, which was once used in vaccines, but it is "considered quackery by almost all doctors." Gardner correctly asserts that the notion that vaccines cause autism "has been thoroughly discredited by dozens of studies [that's even more true today]. . . . Winfrey buys the myth hook, line, and sinker. She has promoted McCarthy's absurd views on numerous shows" and even had her own production company sign up McCarthy for her own talk show. He continues, "McCarthy's crusade is likely to result in needless deaths of children who succumb to diseases that could have been prevented by vaccinations."[43]

Gardner also condemns Winfrey for promoting actress Suzanne Somers's pleas to use only "natural products" and hawking a "bewildering variety of vitamin supplements" and other useless claims. He notes, "Winfrey's enthusiasm for Sommers's wild medical opinions is boundless." He also describes how she promoted Dr. Oz and his uncritical touting of high-priced food supplements, acupuncture, and other alternative medical remedies. Oprah also promoted a host of New Age books, including—in Gardner's words—the "monumental idiocy," *The Secret*, which "teaches that the universe consists of a vibrating energy that can be tapped into with positive thoughts, allowing you to obtain *anything* you desire—happiness, love, and of course fabulous wealth."[44] Thanks to Winfrey's tireless promotion of the book, *The Secret* sold more than seven million copies in the United States alone. Gardner ends his column by quoting David Gorski, a physician at Wayne State University School of Medicine and another well-informed critic of alternative medicine: "The bottom line is that, when it comes to medicine and science, [Winfrey] is a force for ill."[45]

How all of Winfrey's many public programs and activities balance out in the end can be left to history. I'm sure the net effect is positive. But just think how much stronger her legacy would be had Winfrey, with her worldwide audience of devoted fans and admirers, chosen to embrace and promote *good* science and critical thinking about medical and health matters instead of medical pseudoscience and dubious claims.

The actress Gwyneth Paltrow has succeeded Oprah and Oz as our most powerful purveyor medical-related nonsense. (Oprah was one of Paltrow's supporters, as well.) Paltrow's variety of the trade goes even beyond that, seemingly reaching new heights of outrageousness and silliness—engaging in self-promotion and product promotion that must be loved by her bankers and accountants. If she ever had any grasp of science or even logic, she long ago cast that to the winds.

Her company, her enterprise, her empire is the improbably named Goop. (A branding advisor told the actress that all successful internet companies have two *O*s in their names. She added those to her initials.[46]) Paltrow launched Goop in 2008 "from her kitchen as a homespun weekly newsletter." To say that it's "grown a lot since then" (the words of her website) is an understatement. The site claims, "We operate from a place of curiosity and nonjudgment, and we start hard conversations, crack open taboos, and look for connection and resonance everywhere we can find it."[47]

Nonjudgment is indeed key for her. Science or critical thinking need not apply. Any judgment involved must be all about business (whatever sells, with vigorous promotion). Born in 1972, Paltrow underwent her first master cleanse in 1999 and tried a macrobiotic diet for the first time. Five years later, she underwent a cupping, another fad pseudoscience, causing quite a stir. In 2009, her new company created the first Goop detox. It jumped into e-commerce in 2014. She's since written cookbooks and developed her own line of clothing and skin-care products, and in 2015 for the first time mentioned on her website "vaginal steaming." After moving from New York to Los Angeles, the company introduced a glow peel, a pore-refining tonic, a "body luminizer," and other skin-care products. Soon after launching her podcast, it reached 30 million viewers.

Goop's products include a vagina-scented candle for $75, a vampire-repellent spray for $27, and a Goop jade vaginal egg for $66. Yes, a jade egg to place in one's vagina! So what's wrong? The total absence of science for any of it. Goop claims its vaginal eggs could balance hormones and healing through crystals. In September 2018, Goop settled a $145,000 lawsuit for its "misleading" claims about that. Ten counties in California claimed the company had made unscientific claims about the health benefits of its jade egg and rose quartz egg, which the company advised inserting into the vagina "to increase sexual energy and pleasure." A third product named in the suit is the Inner Judge Flower Essence Blend, which the company claimed would help prevent depression.[48]

One online reporter says, "Goop embodies the extremes of wellness culture and pseudoscience permeating into mainstream adoption. But there's a reason why it has received legitimacy and notoriety: Gwyneth Paltrow."[49] Orange County District Attorney Tony Rackauckas, representing one of the ten counties involved in the lawsuit, said in a statement,

"It's important to hold companies accountable for unsubstantiated claims, especially when the claims have the potential to affect women's health."[50]

According to an online report on Goop's claims, Goop asserts that "its Body Vibes sticker used 'NASA spacesuit material' to 'rebalance the energy frequency in our bodies.' The bad thing is, NASA denied this. Two representatives stated there were zero merits to the statement."[51] Goop retracted the claim, saying it was made in error. It then released a statement trying to defend its actions.

The report continues, "When a company spreads misinformation and promises health benefits, it becomes an ethical question. . . Many times when we hear about insane products, we laugh it off. But Goop and Gwyneth Paltrow provide legitimacy, and it's hazardous. Moreover, Goop actively propels those pseudoscience products into the mainstream."[52]

In 2021, when the Center for Inquiry, the umbrella organization that houses the *Skeptical Inquirer*, announced its first "Full of Bull" competition, "Goop crapola seller" Gwyneth Paltrow was one of the six nominees. She was the only woman and the only representative of the alternative medicine industry. (Unfortunately, she didn't win, but I think that's mainly because Trump huckster Rudy Giuliani was in the competition; he easily made off with the top prize.) Robyn Blumner, CFI's president and CEO, refers to Paltrow's "lifestyle-branded absurdities she foists on womenkind" through her Goop empire. She says Paltrow shared much in common with another nominee, Jerry Falwell Jr.: "They both pander dangerous nonsense that gives people false hope and lures them into parting with their hard-earned money."[53]

For some additional common sense about such things (whew!), let us turn once again to my SkepDoc colleague, physician Harriet Hall. In her first *Reality Is the Best Medicine* column for the *Skeptical Inquirer*, Hall chose to write a bit about Paltrow. She titled her column "The Care and Feeding of the Vagina." First of all, Hall says, the vagina doesn't need cleansing or other procedures. It cleans itself, so Paltrow's steaming of the vagina is useless. Hall explains, "Experts have spoken out against it; there are no health benefits and a real possibility of harm." She places Paltrow's jade eggs under the heading "Jade Eggs and Other Quackery." Hall notes that earlier purveyors of pseudoscience have advocated herbal detox pearls, Japanese vaginal-tightening sticks, and ground-up galls (abnormal growths that form when wasps lay larvae in branches). And many, many gels, creams, pills, and sprays. She asserts, "There's not a scrap of evidence that they do anything good, and they're quite likely to do something bad, such as drying out the vaginal mucosa and causing an infection."[54]

Hall calls Paltrow's jade eggs one of the most notorious: "Yes, she wants you to stick a $66 rock up there." Paltrow's claims about that include that it will help "breathe passion" if you use "Yoni Breathing" in conjunction with the jade egg to increase your life-force energy. Paltrow says the jade eggs are an ancient Chinese practice used by concubines, and they "will detox, improve your sex life, balance your menstrual cycle, and intensify feminine energy." Will the egg really do all that? Hall proclaims, "No, it won't. And the porous rock might harbor bacteria." She notes that noted gynecologist and science activist Jen Gunther—whom I've heard speak at several CSICOP conferences—has "published an open letter to Paltrow explaining why her vaginal jade eggs are a bad idea."[55]

CBS News Sunday Morning, a popular and respected television news show that my wife and I try to see every week, did a seven-minute feature on Paltrow on September 25, 2022, which we saw live. Paltrow was turning fifty that week. Parts of the segment gushed about Goop, but the show stated that her $250 million business is controversial. It noted that some doctors and scientists have criticized her promotion of unproven claims. Correspondent Tracy Smith asked Paltrow, "What do you say to people who say that you're promoting pseudoscience?" Paltrow, an attractive woman and a natural on TV, was all innocence. She smiled demurely. "I genuinely don't understand where that comes from because we don't do that," Paltrow said. "We've never done that."[56]

She said some of her products are based on modalities that have "no scientific backing but have worked in India for thousands of years or worked in China," thus committing the fallacy of appeal to ancient wisdom, the idea that because an idea has been around for thousands of years, it must be valid. Paltrow continued, "So I think it was a way of taking shots at us. But there's nothing that we talk about that's actually *that* wacky."

Correspondent Smith jumped on that: "Some of it's maybe a *little* wacky?" she suggested.

"Okay, like what?" asked Paltrow.

"Maybe the egg?"

A narrator pointed out that Goop's $66 jade egg for insertion in the vagina resulted in an investigation by a California task force for false claims in 2018, and Goop settled the case for $145,000. They admitted no wrongdoing but offered full refunds and tweaked the product description—"and they're still selling the egg online." Newspaper headlines about these legal judgments against her were displayed on the screen.

Paltrow wasn't deterred. Her answer was almost chipper: "There's a whole industry now around strengthening your pelvic floor. We were just early."[57] You can see how Paltrow is such an appealing promoter on television. She seems likable. When the segment ended, my wife, well aware of all our concerns about Paltrow's bogus products, commented simply, "I am sure she is a nice person."

You don't have to be a physician to become a critic of the SCAM claims that celebrities love to hawk. Timothy Caulfield is a professor of health law and policy at the University of Alberta in Canada. That sounds nice and official. You might picture him in a stuffy armchair. Don't. He is fascinated with celebrities and their effects on our culture, and he has an energy and a lively informality about him that goes along with that common obsession.

Armed with a year-long subscription to *People*, Caulfield immersed himself in celebrity culture. He did his own firsthand investigations into a bunch of their diets; their antiaging products; and their activities, such as searching for fame by auditioning for *American Idol*. He didn't just study these things from his academic roost; he undertook them himself. "Being near the famous does create an odd, unquantifiable electricity," he says.[58]

He even took the twenty-one-day Clean Cleanse that Gwyneth Paltrow promotes. He went to the Santa Monica offices of the two men who run the cleanse program and signed up for the cleanse. It just about killed him. First, he had to give up coffee. Coffee! He loved coffee. It was central to his existence. "You must give your adrenals a rest," they told him.[59]

But he was determined. He went on the strict cleanse diet. Three days in, he found that his family was "ready to cleanse me from their lives."[60] They questioned his sanity, found his eating habits revolting, and told him his breath stinks. Not a great start. But he persisted. No dairy products. No wheat or gluten. No sugar. No alcohol. And so on. He was miserable. But he persisted. He heard a lot about being detoxified.

The cleanse had absolutely no impact. The Gwyneth cleanse, extreme as it is, did not change the makeup of the bacteria in his colon or "reset" his gut. His fecal matter was just as healthy as before. He survived the ordeal, but for what gain? Well, one result was he could write a lively account about it in his 2015 book, which has one of the greatest titles of all time: *Is Gwyneth Paltrow Wrong about Everything?*[61] It's a rhetorical question.

The idea of cleansing and detoxifying is one of the overarching principles of alternative medical proponents. It is largely driven by celebrity endorsements. It is hugely popular. It rests on the idea that we live in such a toxic environment that toxins constantly build up in our body, which has no defenses against them, and they must regularly be removed artificially. We need to be scrubbed clean. Repeatedly. This way, we will promote natural healing, reduce stress on our adrenal glands, and "reset."

"Let's start with the central idea," says Caulfield in the book. "Do our bodies need to be detoxified and will cleansing do the trick? No and no." He and many other health care professionals have pointed this out, to little avail. "There is absolutely no evidence to support the idea that we need to detoxify our bodies in the manner suggested by the cleansing industry. . . . The idea of detoxing is faulty on so many levels that it borders on the absurd." He, like physicians before him, points out that we have organs that do that for us: the kidneys, liver, skin, and colon among them. He explains,

> When you pee, you are detoxifying. Toxins don't build up, waiting to be cleansed by supplements and special foods. . . .
>
> Second, and more important, there is no evidence that the products and diets sold by the cleansing industry—whether juices, supplements, or specific diet regimes—do anything to help clear toxins, parasites, or bad karma in a manner beneficial to our health. There does not appear to be even a single scientific study to back up the theory behind this massive industry.[62]

Paltrow is hardly the only target of Caulfield's quest. Caulfield explains, "Whether we like it or not, celebrity culture has a profound impact on our world." One big area of impact is the promotion of alternative health claims, remedies, and treatments. Caulfield's firsthand research analyzes and debunks the messages and promises we all hear from celebrities, "be they about health, diet, beauty, our ambitions, or what is supposed to make us happy."[63]

What is to be done? SCAM treatments, alternative medicine, medical pseudoscience, pseudomedicine—whatever we choose to call it—remain popular. They are all backed by enormous industries and promotional campaigns that overwhelm any science to the contrary about their claims and assertions.

Perhaps it is up to us. We need to tune up and turn up our BS detectors. We need to activate that natural skepticism we are born with. We need to turn away from celebrities and all others who hawk dubious and untested products. We need to rely more on real doctors and real medical advice—you know, the responsible kind.

"As consumers, we have certain responsibilities," says Paul Offit in the conclusion to his book *Do You Believe in Magic? The Sense and Nonsense of Alternative Medicine.* He goes on,

> If we're going to make decisions about our health, we need to make sure we're not influenced by the wrong things—specifically, that we don't give alternative medicine a free pass because we're fed up with conventional medicine; or buy products because we're seduced by marketing terms such as *natural*, *organic*, and *antioxidant*; or give undeserved credence to celebrities; . . . or fall prey to healers whose charisma obscures the fact that their therapies are bogus.
>
> Making decisions about our health is an awesome responsibility. If we're going to do it, we need to take it seriously. Otherwise, we will violate the most basic principle of medicine. First, do no harm.[64]

We need instead to turn to real medical science and its abundant peer-reviewed research studies that are designed to determine what really works and what doesn't and what is safe and effective and what isn't. And we need to turn to the doctors and health care providers who draw on that reliable, science-based information in their efforts to treat their patients. If not, we'll continue to be deceived by those who care less about facts, truth, and evidence than about advancing their fame and lining their own pockets—all to the detriment of our own health and welfare.

CHAPTER SEVEN
THE VALUES OF SCIENCE

I have attended scientific meetings all my professional life, and one aspect of almost all the talks I've heard still surprises and impresses me. The speakers typically lace their talks with phrases like this:

I can't be sure.
The data don't allow a firm conclusion one way or another.
We weren't able to get a statistically significant measure.
I am not certain.
Until we can do a fuller study we won't know for sure.
I think it means . . . , but there are other interpretations.

For those who have a stereotypical view of science as a set of strong assertions about the world or of scientists as inflexible or dogmatic, the language of a typical scientific presentation (once you get past any unfamiliar scientific language) would quickly disabuse you of that notion. Such presentations, if you listen carefully, are filled with nuance and uncertainty.

To check my memory about such honest and almost endearing expressions of uncertainty, I've gone back and looked at some papers and abstracts given out at scientific meetings I've attended. Here are some phrases that stick out to me in this regard:

"However, observational evidence was and still is too scarce to decide."

"Possible culprits are . . ."
"On the other hand . . ."
". . . appears to be very high but is difficult to measure."
"In addition, notable exceptions to this position will also be discussed."
"Critical questions . . . remain unanswered."
"The preliminary result . . . has theorists reexamining their models."
"The mechanism remains a matter of debate."
"Meanwhile, uncertainty and confusion abound."
"Although in generally early stages, the results so far are encouraging."
"Such limitations severely limit its use for certain applications."
"A reappraisal of data analyses is required."
"Although in the ideal situation, the evidence will all point to the same ultimate conclusion, in practice, such neat convergence is rare."
"But only recently have we begun to incorporate the data from other sciences."
"Scientists must solve measurement challenges of an ever-increasing complexity."

Embracing Uncertainty

Does it seem paradoxical that such persistent expressions of uncertainty prevail in scientific talks? It shouldn't. Yes, such presentations are expected to give us preliminary new insights into the nature of the world, and science ultimately does. Yet uncertainty is part of the fabric and texture of science. It can't be brushed off and swept away. It has to be accounted for. Scientists deal with it constantly; in fact, they measure it. Scientific data is usually accompanied by quantitative measures of the uncertainty.

"The scientist has a lot of experience with ignorance and doubt and uncertainty, and this experience is of very great importance," Richard Feynman, the Nobel laureate physicist, said. "We have found it of paramount importance that in order to progress we must recognize our ignorance and leave room for doubt. Scientific knowledge is a body of statements of varying degrees of certainty—some most unsure, some nearly sure, but none *absolutely* certain."[1]

What ultimately happens is that over time, the uncertainties surrounding a given research finding lead to deep thinking and better questions. These lead to better hypotheses and then to new and better studies; those in turn lead to better observations and evidence and sharper and more reliable insights. The uncertainties shrink, and more or more scientists become willing to accept the findings as provisionally a good description of that aspect of nature. Or that doesn't happen, and the observations fade away. So the language of uncertainty, to my mind, is one of the most underappreciated elements of good science. Nature is astonishingly complex, and some humility when facing the realities of its mysteries is healthy. All scientists realize this.

Pseudoscience, in contrast, gives us something else. While scientists embrace uncertainty, pseudoscientists abhor it. To pseudoscientists, uncertainty is anathema. It is a sign of weakness. To them, the world is a certain way (or a set of certain ways), and no inconvenient or contradictory evidence from those pesky scientists shall be admitted. Bold claims and bald statements with little if any supporting scientific evidence are the norm. There *is* an afterlife, ghosts exist, mystical forces connect us all, psychics can foretell the future, concocted and untested remedies can cure cancer, UFOs are scurrying about our skies with alien crews either (a) watching over us like gods or (b) abducting unsuspecting humans and subjecting them to humiliating sexual examinations, the forests are filled with undiscovered nonhuman bipeds, the planets' positions correlate with events in our everyday lives, and we can channel 30,000-year-old entities for wisdom about life in the twenty-first century. Isaac Asimov once wrote pithily about this contrast: "Inspect every piece of pseudoscience and you will find a security blanket, a thumb to suck, a skirt to hold. What have we to offer in exchange? Uncertainty! Insecurity!"[2] The opposite ways that science and pseudoscience deal with *uncertainty* is just one of the identifying characteristics that separate them.

Another telling characteristic is *willingness to change*. Good scientists eventually revise their views about any matter if better evidence arises. That is part of the process of science. Scientific research is, after all, about finding new and better evidence, and as that happens, scientists take note of it and gradually begin to change their views accordingly. Not everyone does so at once, of course, and some may not in a given case. Scientists are human. We all want to hold onto our pet ideas and theories; scientists are no different. A few reject all new evidence that casts doubt on their ideas and defend them vociferously. That's okay. That's also how science advances. Without fierce defenders of each particular hypothesis, new research might be too easily accepted. It must pass a barrage of tests and challenges. It is only after this very human process plays out, sometimes over many years, that a new finding becomes provisionally accepted new knowledge.

In one of my earlier books, I wrote about a scientist whose research I was reporting on. It was an exciting new discovery in the field of archaeo-astronomy. He was a good scientist, and he knew that his discovery, if true, must survive challenges from other scientists who might have different interpretations. He voluntarily gave me the name of several of his scientific critics so I might contact them and get their perspective.[3]

Richard Dawkins tells of a formative influence on him as an undergraduate. An elder statesman in the Oxford Zoology Department listened as an American visiting scientist just publicly disproved his favorite theory. Dawkins says the old man strode to the front of the lecture hall, shook the American's hand warmly, and declared, "My dear fellow, I wish to thank you. I have been wrong all these years."[4]

These two examples demonstrate a positive scientific attitude. They show a willingness to subject one's own ideas and research to examination by others. They show a realization that no one need take a new finding seriously unless it survives other knowledgeable scientists' critical examinations.

To most pseudoscientists, changing your views in the light of contrary evidence is an alien idea. They want certainty. They often claim certainty. Such expressions of absolute certainty are a sure sign of pseudoscientific thinking. A more cautious appraisal, admitting that the evidence may not yet be conclusive and perhaps suggesting how it can be bolstered, if it indeed is correct, is one mark of good science.

A related quality of science might be called *humility*. Faced with the awesome mysteries of nature and the incomprehensible scales of the macrouniverse and microuniverse, almost beyond any intuitive human understanding, knowledge seekers try to solve some of those mysteries, one at a time. They peer behind the curtain of our ignorance, only to see further layers of the unknown. How can science not be humble in this noble quest? As the American philosopher Rebecca Goldstein says, "Science, the enterprise, is the embodiment of humility. It makes a method out of humility, directing its ingenuity to trying to get nature to correct us when we get it wrong." In science, she emphasizes, "nothing is immune to revision except the very normative standards of rationality that allow us to do science in the first place."[5] My colleague anthropologist Eugenie Scott of the National Center for Science Education states, "'I don't know yet' is the most common statement in science." As she notes, that statement is "rarely used in pseudoscience."[6]

So embracing *uncertainty*, a *willingness to change*, and a certain *humility* in facing the difficulty posed by nature's complications are three crucial, related characteristics of science that I think are little understood and underappreciated, so much so that I thought it important to begin this chapter on the values of science with them. They are little seen in the pseudosciences.

Probably *the* most important value of science is a *respect for truth*. This intense desire to find the truth about nature is so interwoven into the fabric of science that it is inseparable. Without it, science would be something else entirely. (I am sure you can think of many common endeavors where truth is not highly valued.) It is difficult to talk about this quality without sounding sanctimonious. That's probably why scientists seldom discuss it with the public. But I think it is important for everyone to understand how powerful that aspiration is in science and how powerful its realization is to science. The importance of truth to all of science is one of the strongest reasons I admire science so much. Not uncritically so—scientists are human and have their flaws and shortcomings. But to most good scientists, the ideal of truth shines brightly like a beacon. If they have reverence for anything, it is to the truth.

In his book the *Common Sense of Science*, the biologist/mathematician and great humanist Jacob Bronowski puts it this way:

> For whatever else may be held against science, this cannot be denied, that it takes for ultimate judgment one criterion alone, that it shall be truthful. . . . T. H. Huxley was an agnostic, [nineteenth-century geometer and philosopher William] Clifford was an atheist, and I know at least one great mathematician who is a scoundrel. Yet all of them rest their scientific faith in an uncompromising adherence to truth, and the irresistible urge to discover it.[7]

Wherever the evidence leads, that's where science eventually goes. I think this is one of the most positive, admirable qualities of science, and I wish more people shared it. Scientists deeply and unreservedly value truth—truthful statements, truthful assessments of evidence, truthful evaluations of one's own natural biases and prejudices. A correlative individual value is *honesty*. Honesty and integrity are highly valued among scientists. People who violate this norm in significant ways—say, by presenting fraudulent results—are severely punished by loss of jobs, loss of reputation, and loss of all standing in science.

"Truth is the most treasured value the scientific community possesses," says the biologist and philosopher of science Gerard Verschuuren in his book *Life Scientists*. "Complete intellectual honesty is a first essential in scientific work. . . . In science, lies are never allowed. The truth is a public interest, to be protected from sectional interests that go in the opposite direction."[8] The quest for truth in science is aspirational yet never absolutely fulfilled. Mathematics and logic may lay claim to truth, but the processes of science lead at best to closer and closer approximations of truth. And new evidence can always move the bar farther away. Scientists go where the evidence goes.

A related value—Verschuuren puts it at the same level—is *objectivity*. Objectivity is also an ideal that can never be achieved. But the processes of science are designed to minimize scientists' innate subjectivity. This is accomplished via a number of methods, criteria, and tests that enable other informed members of the scientific community to repeat, check, and criticize scientific results. Individual scientists may not be wholly objective, but the processes of science do eventually result in a more objective and therefore more accurate view of nature.[9] Objectivity also leads to acknowledging and seeking out alternative or even contrary ideas and information, testing one's own explanations against competing ones, and then revising accordingly.

Passions motivate scientists during the discovery phase of their research, and that should always be the case. That passion to know, to find something new about nature, is essential to discovery. It must not be discouraged. But objectivity, or disinterestedness, comes into play in the evaluative phases, where clear judgments must be made assessing the reliability of a result. "The value of objectivity and its derivative values, intersubjectivity and neutrality, are as important elements in doing research as the value of veracity and academic freedom," Verschuuren writes. "All of these values are highly esteemed by the scientific community. . . . Science became a highly valued activity because it is understood to be a disinterested and unrestricted search for objective truth."[10]

There is still another value of science not fully appreciated by the public: its *collaborative, cooperative nature*. We refer to the scientific community, but there are many smaller communities of scientists in each subfield who share common interests, methods, and values. They know each other's work and can be useful to each other in shaping new research directions and deciding when a new idea has potential merit. Furthermore, everyone realizes that all science proceeds from previous scientific work. ("On the shoulders of giants," as Sir Isaac Newton said; he just had broader scientific shoulders than most.) The lone genius is today a stereotype that falls to the reality: many scientists and technicians with a variety of skills relevant to a given problem who work cooperatively to gain some slight new foothold on a potential scientific advance.

When a team of scientists announced they had managed to obtain the first-ever image of a black hole—the supermassive black hole in the center of a giant galaxy, M87 (see chapter 1)—this epic achievement was the result of a massive international collaboration. It had been envisioned a decade and a half earlier and then required enormous planning and coordination (and some new technological advances) to carry it off. Hundreds if not thousands of scientists, computer engineers, and administrators at observing sites, laboratories, universities, and other research institutions worldwide were involved. Ten radio telescopes on four continents (including one at the South Pole), to effectively create a telescope the diameter of the Earth, collected the data over a prearranged three-day period. The massive quantities of data were then stored on hard drives and eventually flown to central locations and compared, calibrated, and analyzed. Publication of the results required six separate papers in *Astrophysical Journal Letters* (April 10, 2019). I downloaded, printed, and read the seventeen-page summary paper, "First M87 Event Horizon Telescope Results. I. The Shadow of the Supermassive Black Hole," by the Event Horizon Telescope Collaboration.[11] (The other five papers describe the Earth-diameter-observing array, data-reduction methods, imaging of M87, theoretical models, and how the mass of the black hole was estimated to be equivalent to 6.5 billion Suns.)

The list of around two hundred authors of just this first paper is so long that it had to be placed at the end and required four very small-type, single-spaced pages. The authors are listed alphabetically. Those from the United States alone represent, by my count, forty-three university departments, labs, and institutes; other observatories and research institutes; and several private entities from across the country. But it truly was an international team. The project head, Heino Falcke, is a German-born astrophysicist working in the Netherlands. Other authors came from research centers in Spain, the Netherlands, France, Canada, South Korea, China, Sweden, Japan, Italy, Germany, South Africa, Mexico, Russia, Taiwan, Chile, Finland, the United Kingdom, and Malaysia.

Think of planning and coordinating such an enormous project—and then having it all come to fruition in one remarkable image of a black hole that made all the international and national newscasts and stimulated discussion (mostly serious but some lighthearted) and emotional reactions of amazement and awe on a scale that few scientific accomplishments ever achieve. Planetary exploration programs observing the distant reaches of the solar system require much of the same long-term planning and coordination (and have some of the same public impact). So, too, does much modern genetics and gene-mapping research, contemporary biomedical research, and many other projects at the frontiers of biological science.

In 2020, an enormous group of 230 biological scientists coauthored a paper in *Science* on the long-term thermal sensitivity of Earth's tropical forests, a topic crucial to understand as the planet warms. Accomplishing this task required assembling a network of 590 permanent plots across tropical lowland forests in 24 countries in South America, Africa, Asia, and Australia. Data were collected and analyzed from more than a half-million trees. The researchers represented, by my count, 178 different research institutions in countries around the world, not to mention scores of funding agencies and supporting foundations in Great Britain, the United States, Brazil, and elsewhere. Imagine the coordination this program required! The top two-thirds of the opening page of the published article is devoted just to listing all the authors' names, and the listing of authors' affiliations at the end of the article takes up nearly four columns—all in very small print.[12]

"There's this notion that science is a solitary life, that you wear a white coat, go into a basement lab, work for years and occasionally you come out to see the light and be dusted off," California Institute of Technology President Thomas Rosenbaum, an experimental physicist, said in a 2019 interview. "It is in fact a remarkably interactive experience to be a scientist. You basically run a small business if you run a laboratory. You intersect with colleagues around the world. You speak the language that cuts across international boundaries. You ask questions that are compelling and obsessing."[13]

Not only is the discovery process collaborative and communal, but so, too, is everything that follows. Scientists usually share and discuss their preliminary results with colleagues and get their feedback before proceeding further. When they write up their final results, they share drafts with others and get initial criticisms or comments before revising. Once the written paper is submitted to a journal, a process of initial editorial review and then formal peer review follows. This can be excruciating. Sometimes the criticisms are withering. They may even be unfair, petty, or off base. But the process cuts down on publication of unsupported findings or poorly thought-out research. In many cases, it may suggest good ways to improve the research paper and make it more broadly reliable and useful.

Now what do we see in the fields of pseudoscience? Just the opposite. Many pseudoscientists work in isolation. They may be brilliant and knowledgeable, even insightful. But without the continual interaction with colleagues in a community of scientists who can help keep them on a reasonable scientific track, they can easily diverge into deeply unproductive and unscientific paths. And when that starts happening, it can be a spiral downward into nowhere. The "hermit scientist" is often a pseudoscientist.

(I suspect in this new digital age, the isolation of pseudoscientists is a problem of a different character from what it used to be before the internet. Now they can easily reach out and find people who *believe* as they do. But if they only self-select admirers and information from those who share their obsessions, they are no better off than before, perhaps even worse. If they are deluded, they may have their delusions reinforced and deepened.)

As for *truth*, in many areas of pseudoscience, it unfortunately is not highly valued. You can see that in the way pseudoscientists usually ignore all previous science that falsifies their claims, misstate or distort the science they do acknowledge, and make no effort to have their assertions checked and tested independently. Truth is not what they are after. Uncritical acceptance is what they want. They seek only information and arguments that might support what they assert. They are the poster children for the pitfalls of *confirmation bias.*

As for objectivity, with pseudoscientists, objectivity is not even an objective! Subjectivity reigns. Personal feelings, wishes, hopes, and opinions can all be seen in abundant display. These are all fine and understandable human traits, but they are not what you want when evaluating the likelihood that a proclaimed discovery, concept, remedy, or product is valid, true, or works as hoped. Here you want scientific objectivity, not personal subjectivity.

So far in this chapter, I have emphasized the *values* of science. That is because I think these values are central to our understanding of why science itself should be valued and because these values are little understood and appreciated by nonscientists. That is especially so in this age of distrust, when science is often considered or misleadingly portrayed as just another set of opinions about the world.

There is another important value, already alluded to, that is ingrained within science but must be supported from the outside for science to thrive: *freedom of inquiry*. All scientists and scholars value freedom of inquiry—the right to study and investigate virtually any topic or idea, to openly discuss their results and their significance, and to publish or otherwise distribute them to the wider community. But this value of inquiry and openness also depends on its being appreciated and shared by the wider public. Science, innovation, and creativity (and this is true of the arts as well as the sciences) thrive in a cultural, political, social, and intellectual environment of openness and active ferment, in which new ideas and new ways of thinking are welcomed and appreciated. In contrast, when independent thinking, new ideas, scholarship, and innovations are suppressed, dogma and authoritarianism easily take hold. Progress in improving the human condition slows, stops, or even reverses. We have seen this many times in human history, and we see it happening now in many parts of the world.

The Beginnings of Science

The eras of great intellectual and scientific insights started with the Ionian Greeks of the fifth and sixth centuries BCE. This is when rationalism and the two big trunks of rationalism—science and philosophy—were born, a time aptly described by historian Richard Berthold as an "electrifying moment in human history." Here arose a "purely scientific inquiry into the nature of man and the universe."[14] It was without religion or dogma, and it sought consistent and logical generalizations about nature independent of any supernatural or godly interpretations. These early Greeks also became infected with the idea of progress—another entirely new idea at that time. (The ancient Egyptians had no such concept.)

Here, too, was the beginning of skepticism—another important attribute of science, in which skeptical scrutiny of all ideas is not just tolerated but also considered an integral attribute of intellectual thinking. They focused their inquiries on purely impersonal forces, another radical idea for the time. Here they broke free of a mythmaking universe that had the gods constantly meddling in human affairs.[15] We owe an enormous debt to these early Greek thinkers.

We skip ahead almost two millennia. Leonardo da Vinci (1452–1519), an extraordinarily creative master in both the arts and sciences, was energized by some of that same spirit of the great Greek thinkers. He had an amazingly independent mind. He dedicated himself to determining, through detailed experiment and observation (whether it be the origin of landforms or the anatomy and physiology of the human body), what was really true, rather than just accepting assumed or passed-down knowledge. Through his remarkable studies, he gained insights that weren't fully accepted for several centuries.

Copernicus offered his then radical revision of the established view of the universe in 1543, contending (correctly, it turned out) that the Earth is not the center of the universe but in fact, along with the other planets, is in orbit about the Sun. Still, it was nearly another half-century or so before scientific thinking with systematic observation and experimental testing began to reshape human culture in western Europe. Galileo began to devise all manner of experiments that questioned and disproved many Aristotelian ideas about mechanics and physics, and he sought a quantitative understanding of nature. He began to support Copernicus's ideas.

Through his new telescope, starting in 1609, he observed the heavens systematically and saw mountains on our Moon; four moons moving around Jupiter; sunspots revealing the Sun's twenty-eight-day rotation; "horns" that turned out to be the rings of Saturn; and the stars in their multitudes, still points of light in his telescope. All this led him to discern that the universe was much different and much bigger than had been supposed. A scientific revolution was underway.

The Nature of Science

In his *Novum Organum* (1620), the first clear call for an age of reason, Francis Bacon describes the actual method of modern science: "The true method of experience first lights the candle [by hypothesis], and then by means of the candle shows the way, commencing as it does with experience duly ordered . . . and from it educing axioms, and from established axioms again new experiments. . . . Experiment itself shall judge."[16]

Bacon's influence and passion for the use of new knowledge to better human welfare stimulated the thinkers and scientists of his time and helped create a cultural environment that valued science and discovery. Not a scientist himself but a keen observer and philosopher of science, Bacon was the most powerful and influential intellect of his time. Will and Ariel Durant exclaim, "How refreshing, after centuries of minds imprisoned in their roots or caught in the webs of their own wishful weaving, to come upon a man who loved the sharp tang of fact, the vitalizing air of seeking and finding, the zest of casting lines of doubt into the deepest pools of ignorance, superstition, and fear!"[17]

When the Royal Society was founded in London in 1660, Bacon was acknowledged as its inspiration. Say the Durants, "He repudiated the reliance upon traditions and authorities; he required rational and natural explanations instead of emotional presumptions, supernatural interventions, and popular mythology. He raised a banner for all the sciences, and drew to it the most eager minds of the succeeding centuries."[18] The age of reason and the age of discovery were both well underway. Science in something close to its modern form had been born.

The British Nobel laureate biologist Peter Medawar (1915–1987) was one of our most eloquent and insightful observers of the nature of science. (In addition to his scientific work, he wrote regularly for London and

New York newspaper literary supplements.) He once lamented that the public had "two completely different conceptions of science," embodying two different valuations of scientific life and of the purpose of scientific enquiry. In the first, science is seen, above all else, as an "imaginative and exploratory activity, and the scientist is a man taking part in a great intellectual adventure." The alternative conception, he says, ran something like this: "The scientist is pre-eminently a man who requires evidence before he delivers an opinion, and when it comes to evidence he is hard to please. Imagination is a catalyst merely; it can speed thought but cannot start it or give it direction."[19]

Let's leave aside for a moment the facts that his statement of these two opposed conceptions leaves women scientists out entirely. We can attribute that to the time in which he was writing and the fact that not only is he admittedly exaggerating the contrasts between these two conceptions but also that he is calling these ideas "not really thoughts at all, but thought-substitutes."[20] What about these two seemingly diametrically opposed views of science? He writes,

> Inasmuch as these two sets of opinions contradict each other flatly in every particular, it seems hardly possible that they should both be true; but anyone who has actually done or reflected deeply upon scientific research knows that there is in fact a great deal of truth in both of them. For a scientist must be freely imaginative and yet sceptical, creative and yet a critic. There is a sense in which he must be free, but another in which his thought must be very precisely regimented; there is poetry in science, but also a lot of bookkeeping.[21]

This description will resonate with most scientists in one way or another. The first conception is admittedly a somewhat romantic view of science, but the role of imagination in scientific discovery can't be discounted. If nothing else, it motivates scientists. In some cases, it inspires them and leads to new insights. The second view is more worldly or utilitarian. But science is both a creative activity and one requiring immense discernment and critical judgment. How these two broad characteristics balance off one another varies in every instance.

"There is no paradox here," Medawar concludes. "It just so happens that what are usually thought of as two alternative and indeed competing accounts of *one* process of thought are in facts accounts of the *two* successive and complementary episodes of thought that occur in every advance of scientific understanding."[22]

Imagination and critical judgment: Both are an absolute necessity in science. Pseudoscience usually has an abundance of the former but a shortage of the latter. It remains mostly untethered to the real world because it undervalues or totally ignores rigorous, critical examination of the evidence. Einstein is famous for having once said to an interviewer, "Imagination is more important than knowledge." Those words are emblazoned on popular T-shirts and posters. What he meant, and he so explained in his next sentence, is that "knowledge is limited. Imagination circles the world."[23]

But here's something else he said that is at least equally important. He wrote it in a personal letter replying to an American man who had written him: "Imagination is good. But it must always be controlled by the available facts."[24] Now that is equally succinct. It could also fit on a T-shirt, but Einstein's cautionary words, as pertinent and important today as they are, don't have quite the same appeal to the masses as his first, better-known quote.

Feynman, the Nobel laureate physicist quoted earlier in this chapter, says much the same thing but explains it in a bit more detail. In his famous *Lectures on Physics*, Feynman describes how the scientific imagination has to be tethered to the real world: "Our kind of imagination is quite a difficult game. One has to have imagination to think of something that has never been seen before, never been heard of before. At the same time the thoughts are restricted in a straitjacket, so to speak, limited by the conditions that come from our knowledge of the way nature really is."[25]

This again is where pseudoscience runs afoul of scientific thinking. Pseudoscientists and their adherents don't know, understand, or suitably appreciate the limitations nature imposes on the possible. To pseudoscientists, indeed to many people not familiar with the laws of nature, anything that can be imagined may be possible. It is an appealing thought, but scientists know that isn't the case. As Feynman puts it, "whatever we are *allowed* to imagine in science must be *consistent with everything else we know*. . . . We can't allow ourselves to seriously imagine things which are obviously in contradiction to the known laws of nature."[26]

Informed speculation is also a part of science, but that, too, must be tied to reality. And that reality is found via systematic observation and laboratory experiments. One thing that separates the early Greek philosophers, as brilliant as they were, from even the most mundane of natural philosophers today (whom we now call *scientists*, a word not even invented until 1840) is the laboratory. And by laboratory, I mean not just the technological instruments of science that populate scientific labs but also the experimental methods of science that guide the whole enterprise of science. The two together make all the difference.

Experimentation is a defining characteristic of science. Medawar sees it as a *form* of criticism, the critical judgment I earlier mention: "Experimentation *is* criticism."[27] Laboratories are where hypotheses go to die. If you prefer, I can be more generous and say laboratories are where hypotheses go to be tested. But if they don't pass the tests, they must be abandoned. As Max Planck once pungently put it, "experimenters are the shock troops of science."[28]

And this gets us to *theory*. It is an important word in science but one much misunderstood and misused by the public. To most people, it is simply a hunch or a guess. That is one of the great misconceptions about any word used in science. It is so much more than that. Scientists sometimes do use the word in the casual sense like we all do, but there is another sense of the term that is quite formal and meaningful, and here's where the public gets confused. *Theory* in the formal sense refers to a comprehensive explanation of some aspect of nature that is supported by a vast body of evidence. Theories, in this formal sense, are deep *explanations* that apply to a broad range of phenomena. They may integrate many hypotheses and laws.[29]

So you see the problem: The same word is used popularly to mean essentially a guess and used formally in scientific thinking as something with broad explanatory power supported by multiple lines of evidence. Two totally different meanings. In the formal sense, we have all heard phrases like *cell theory*, *germ theory*, *atomic theory*, *gravitational theory*, the *theory of evolution*, and so on. As a result of a series of epochal advances in understanding in the past two centuries, we all now know that living things are composed of cells and that all cells arise from other cells; that infectious diseases are caused by microorganisms; that matter is composed of atoms, with protons and neutrons at their core and electrons orbiting the nuclei; that gravity is the attraction between two masses (such as your body when you stumble and the Earth when you crash down into it); and that species evolve over time through the process of natural selection (even though this theory is still much resisted on religious grounds). These now are comprehensive and incredibly well-supported theories—so much so that probably no purpose is served any longer by calling them *theories*, especially given our capacity for misunderstanding (using the wrong sense of the term).

When people use *theory* in the casual sense, the scientist might instead say *hypothesis* and mean much the same thing. A hypothesis is essentially a tentative assumption or possible explanation proposed to draw out and test its logical and empirical consequences. It is usually limited in scope and applied to a narrow range of phenomena, perhaps even a single observation. Nevertheless, it is much more informed than a guess, and there may already be fairly broad provisional evidence for it. And it may have some predictive power.[30] Science is all about testing hypotheses.

In this talk about aspects of science, I might as well now present a succinct definition of *science*. Here is one I like:

> The use of evidence to construct testable explanations and predictions of natural phenomena, as well as the knowledge generated by this process.[31]

Scientists and philosophers can and do go on at great length about what science *is* (whole bookshelves of books devoted to it), and those books are interesting indeed for people like me to read. But for the purpose of this book, a very succinct statement like this one will do. It is just twenty-two words. Science is both a process for developing an accurate understanding of the natural world and the body of knowledge that results. (The latter is what most people usually mean when they refer to *science*, as in, "*Science says* . . .")

The word *fact*—like *theory*—has several usages, as well. Typically, a *fact* refers to an observation, measurement, or other form of evidence that can be expected to occur in the same way under similar circumstances. But the same National Academy scientific panel points out that scientists also use *fact* in another sense: to refer to a "scientific explanation that has been tested and confirmed so many times that there is no longer a compelling reason to keep testing it."[32] In that sense, biological evolution, for example, is a scientific fact.

To many people science is essentially fact collecting or a collection of facts, but of course, that is just one small, limited aspect of science. Science seeks to explain things about nature, and those explanations and predictions must be coherent and testable and reproducible. They must also generally be framed in such a way that if they are wrong, they may be proved wrong. This is a concept called *falsifiability*. In the philosophy of science, falsifiability is no longer universally accepted as a defining feature of science (I visit it more in a later chapter about philosophers' ideas and debates about science and pseudoscience). But for now, the idea is nevertheless a fairly useful and quick way to expose ideas that are pseudoscientific from those that are scientific. Just ask the question, "If your idea is false, how can that be proved?" Scientists can always answer that question. Pseudoscientists will be flummoxed by it.

I started this chapter with the values of science. They include acceptance of uncertainty, a willingness to change based on better evidence, a certain humility in the face of nature's mysteries, a deep respect for truth, the necessity of objectivity, collaboration and cooperation, openness, and freedom of inquiry. I then gave a glimpse of the beginnings of science and considered some aspects of the nature of science. The *findings* of science (or at least a few of those that seem especially interesting or important) are widely reported and discussed today, and that is a good thing. But if everyone appreciated more the values and nature of science, we'd be much better off. We'd be better able to affirm the value of scientific inquiry and more easily distinguish good science from its pretenders. More characteristics of science (and of pseudoscience) are evident in the following few chapters, beginning with the next topic, "The Demarcation Problem: Philosophers and Pseudoscience."

CHAPTER EIGHT
THE DEMARCATION PROBLEM
Philosophers and Pseudoscience

There is a famous cartoon by Sydney Harris about science and pseudoscience. It shows two sections of a library. The largest by far, probably 80 percent of the space, is labeled "Pseudoscience." The tiny section to its left is designated "Science."[1] The cartoon more or less accurately portrays the regrettable imbalance between pseudoscience and science in the popular arena. But, alas, the distinctions are hardly that obvious or clear.

No one labels themselves a pseudoscientist. No books are designated pseudoscience. It is not a Library of Congress classification. I made that point earlier, and that problem is just the tip of the iceberg. Distinguishing between science and pseudoscience and exploring the boundaries between the two are important topics in the philosophy of science. Scientists, always pragmatists, tend to do this naturally and almost intuitively (perhaps even subconsciously), but philosophers seek a more rigorous, systematic set of principles to establish such boundaries—or to decide if indeed there are such sets of principles or even if there are distinguishable boundaries in the first place.

The twentieth-century philosopher Karl Popper called this quest for what distinguishes science from nonscience and pseudoscience the demarcation problem. He eventually came to think that a solution to it was what he called his *falsifiablility* criterion, which I discuss in the previous chapter. I, too, think falsifiability is a valuable, and even powerful, criterion, but philosophers of science are fairly unanimous now that falsifiability in itself is inadequate as a criterion to make the necessary distinctions. It is valuable but not sufficient.

One eminent philosopher who has tackled the problem is Mario Bunge. Bunge, both a physicist and philosopher, was for decades the Frothingham Professor of Logic and Metaphysics in the Department of Philosophy at McGill University in Montreal. He had a long and prolific career. He wrote seventy books on physics and on the philosophy of science, even on medical philosophy. In 2016, he published his memoirs, *Between Two Worlds: Memoirs of a Philosopher-Scientist*. For such an erudite academic, it is, I found, surprisingly readable. In 2018, he came back with *From a Scientific Point of View*, a clear exposition on the scientific worldview, updated with recent scientific discoveries. Bunge amazingly remained intellectually active, regularly publishing new books, until his ninety-ninth birthday. In late 2019, his colleagues published a massive 827-page tribute to him, *Mario Bunge: A Centenary Festschrift*, with forty-one chapters by scholars from sixteen countries to honor him on his hundredth birthday.[2] He died five months later, on February 24, 2020, at the age of one hundred. An obituary called him "one of the outstanding figures in twentieth-century philosophy of science" and said, "Few had approached the scope, depth, and detail of his contributions to the discipline."[3]

In an important earlier article for nonspecialists titled "What Is Pseudoscience?" Bunge outlines his systematic thinking about pseudoscience. I think it, and a later related article, "The Philosophy behind Pseudoscience," are so sufficiently revealing that I should give some substantive summary for you here. I then get to the views of a number of other philosophers of science.[4]

No single feature, such as empirical success or refutability, distinguishes science from pseudoscience, Bunge argues. All such simplistic attempts to do so have failed. Science is just too complex to be identified by a single trait, and so is pseudoscience. Instead he offers a list of twenty-two typical attitudes and activities of scientists on the one side and of pseudoscientists on the other. I think this list is interesting enough to reproduce in full (see figure 8.1, courtesy of Professor Bunge).

Note that figure 8.1 presents in convenient form many of the traits and characteristics of science and scientists I describe in previous chapters ("admits own ignorance," "welcomes new hypotheses and methods,"

Typical Attitudes and Activities	Scientist			Pseudoscientist		
	Yes	No	Optional	Yes	No	Optional
Admits own ignorance, hence need for more research	x				x	
Finds own field difficult and full of holes	x				x	
Advances by posing and solving new problems	x					
Welcomes new hypotheses and methods	x				x	
Proposes and tries out new hypotheses	x					x
Attempts to find or apply laws	x				x	
Cherishes the unity of science	x				x	
Relies on logic	x					x
Uses mathematics	x					x
Gathers or uses data, particularly quantitative ones	x					x
Looks for counterexamples	x				x	
Invents or applies objective checking procedures	x				x	
Settles disputes by experimentation of computations	x				x	
Falls back consistently on authority		x		x		
Suppresses or distorts unfavorable data		x		x		
Updates own information	x				x	
Seeks critical comments from others	x				x	
Writes papers understandable by anyone		x		x		
Is likely to achieve instant celebrity		x		x		

Figure 8.1. Comparison of Attitudes and Activities of Scientists and Pseudoscientists. *Source*: Mario Bunge, "What Is Pseudoscience?" *Skeptical Inquirer* 9, no. 1 (Fall 1984): 36–46. Reproduced with permission.

"looks for counterexamples," "invents or applies objective checking procedures," "updates own information, "settles disputes by experimentation of computation," etc.). I think it is useful to show these in list form. Bunge realizes not all scientists behave scientifically on every occasion—they are human, after all!—but that is not the point. He says with these attitudes and activities, "we are concerned with norms or ideal behavior."[5]

Also note that Bunge grants that pseudoscientists do sometimes propose new hypotheses, use logic and mathematics, and gather data. But even when they do so, they fall well short of scientific ideals. Often, all that is mere window dressing, part of the pretend aspect of pseudoscience, the mimicking of some aspects of science without fully applying them in any systematic or critical way. He writes, "Pseudoscience is a body of beliefs and practices but seldom a field of active inquiry; it is tradition-bound and dogmatic rather than forward-looking and exploratory. (In this respect it resembles ideology and, in particular, religion.)"[6]

In presenting these kinds of distinctions, is there a danger of turning our backs on prospective good science that is just not yet fully developed? We wouldn't want that to happen, but generally, that's not a problem. The reason is that science itself welcomes new ideas—as long as they are supported by evidence. Bunge puts it this way: "There is always the fear that some gold nuggets may lie hidden in a pile of pseudoscientific rubbish, that the latter may be nothing but protoscience, or emerging science." But because science is usually open to novel ideas and approaches, "if the novelty fails to evolve into a full-fledged component of science at the end of, say, half a century," then it is probably time to cast it aside. Bunge continues, "Indeed, whereas the protosciences advance and end up by becoming sciences, the pseudosciences are stagnant pools on the side of the swift current of scientific research."[7]

Bunge notes that the alchemists were right in believing that "lead can be transmuted into gold. But they were wrong in believing that they would eventually bring about such a transmutation." Why? They "lacked the necessary theory (of nuclear structure), . . . they lacked the necessary tool (particle accelerator), and . . . they were hooked to tradition (in particular the four elements theory)," so they lacked even the possibility of coming up with the appropriate theory. They "put their faith in dogma, in trial and error (rather than in well-designed experiment), and in magical incantation. So the modern discovery of (genuine) transmutation was just a coincidence—the more so since the alchemists rejected atomism."[8]

Likewise, he argues, telepathy might turn out to be a fact after all, but if that were to happen, then thought transmission would become a part of science because it would depend on a physical process. He holds out no such hope for other claims of parapsychology—clairvoyance, precognition, and psychokinesis—"all of which conflict with basic physical laws."[9]

Bunge concludes his examination by lamenting the fact that most scientists ignore pseudoscience or consider it "harmless rubbish" or something for the amusement of the masses; "they are far too busy with their own research to bother with such nonsense. This attitude is most unfortunate." He lists several reasons: "First, superstition, pseudoscience, and antiscience are not rubbish that can be recycled into something useful; they are intellectual viruses that can attack anybody, layman or scientist, to the point of sickening an entire culture and turning it against scientific research." This argument expresses well some of the dangers in chapter 4. Second, he argues that they are "important psychosocial phenomena worth being investigated scientifically and perhaps even used as indicators of the state of the health of a culture." And third, central to this chapter, he gets to the philosophical issue: "Pseudoscience and antiscience are good test cases for any philosophy of science. Indeed, the worth of such philosophy can be gauged by its sensitivity to the differences between science and nonscience, high-grade and low-grade science, and living and dead science."[10]

Sometime after Bunge's article was published, sociologist of science Marcello Truzzi took issue with it. Truzzi was himself a scholar of pseudoscience, but he often took a view of it different from that of most working scientists. He tended to treat issues of this nature more as open, political-style debates, where as long as every viewpoint was expressed, truth was served. Scientists, in assessing differences between pseudoscience and science, tend to resort directly to the scientific evidence and to the principles and laws of nature. Those, they feel, carry far more weight than even the fair-minded debates of opinion that Truzzi favored. Truzzi also tended to be more generous to pseudoscience in another respect, arguing for or emphasizing their possible "protoscience" nature. He liked to propose that while they had shortcomings, they were, at least in some cases, likely to become emerging sciences. Bunge, of course, addresses that in his article, but Truzzi didn't find much he liked in that aspect of the piece or in the general list comparing attitudes and activities of scientists and pseudoscientists.

In a lengthy article, Truzzi devotes more than two full pages to Bunge's examinations of the topic, including Bunge's *Skeptical Inquirer* article "What Is Science?" Some of Truzzi's attitudes about it are subtle. Truzzi says Bunge's list expresses the attitudes and activities "he [Bunge] thinks" can be used in comparing and contrasting science and pseudoscience. Truzzi finds some of Bunge's criteria outlined earlier in his article "far from unequivocal." He notes that Bunge welcomes dissent from within science, what Isaac Asimov called *endoheresies*, "but does not welcome *exoheresies*, unorthodoxies coming from outside of science."[11]

Yet Truzzi does see some value in Bunge's effort: "Despite its problems, Bunge's approach shows insight and promise that might be made more useful if his criteria for science and pseudoscience were made less rigid. The virtue of Bunge's system is that it is more complex than those of most essentialists, yet it may still not be complex enough to cover the practices we need to describe."[12]

Truzzi wasn't fond of Bunge's binary system of two sets, science versus pseudoscience, and instead proposed they both exist along a continuum. Truzzi contends, "Not only are some pseudosciences more scientific than others, but some sciences may also be more pseudoscientific than others." He continues, "This reconstruction builds upon Bunge's acknowledgement that a science may contain pockets of pseudoscience and that the reverse may be true. It also suggests that pseudosciences may vary from one another by containing different mixtures of scientific and pseudoscientific elements."[13]

Here we can see on display Truzzi's attitudes. They can be seen, on the one hand, as admirably nuanced and intellectually open, and on the other, depending on your point of view, as an example of a frustrating unwillingness to make clear judgments based on the evidence. A few years back, I asked Bunge about Truzzi's criticisms. Bunge was pleased I intended to describe his writings on pseudoscience in this book, but as for Truzzi, he answered brusquely: "Sorry, I have no time for Truzzi. . . . He is appallingly uninformed. Also, unwilling to learn."[14]

Truzzi is one of the most enigmatic figures in modern skepticism. He loved arguing such points. In 2017, a book-length collection of extensive and extraordinary correspondence between Truzzi and the great skeptic, author, and science supporter Martin Gardner was published.[15] These remarkable exchanges reveal that these two quite different people debated endlessly the same demarcation problem I discuss here, in hundreds of detailed back-and-forth letters in the 1970s and early 1980s. They show Gardner's increasing frustration with Truzzi in refusing, for example, to label as a pseudoscientist even Immanuel Velikovsky (*Worlds in Collision*), who in the 1960s and 1970s proposed a bizarre series of planetary collisions as accounting for certain stories of biblical miracles. Velikovsky's views were roundly condemned by planetary scientists, but they found much popular appeal.[16]

Some years after his initial article, Bunge was back with what I considered an extremely worthwhile exploration with "The Philosophy behind Pseudoscience." Here Bunge argues, "Every intellectual endeavor, whether authentic or bogus, has an underlying philosophy," even if "most scientists, as well as most pseudoscientists, are unaware that they uphold any philosophical views." He notes that "they dislike being told that they do." Yet he quickly adds that "nobody can help but employ a great number of philosophical concepts, such as those of reality, time, causation, chance, knowledge, and truth."[17] The problem is that the philosophical ideas underlying science—such as the six stated at the end of the previous sentence—differ totally from those behind pseudoscience. In fact, says Bunge, they are "orthogonal" to each other.

When *Skeptical Inquirer* published this article, I wrote in an editorial that, to me, Bunge's argument "seems powerfully explanatory. . . . I think it provides a comprehensive, explanatory, intellectual, philosophical framework" for virtually every issue that science-minded people concerned about pseudoscience address.[18] A scientific investigation of a domain of facts, Bunge writes, assumes that these are "material, lawful, and scrutable, as opposed to immaterial (in particular supernatural), lawless, or inscrutable." Furthermore, it assumes that the investigation is based on a body of previous scientific findings, and it is done with the main aim of describing and explaining the facts in question with the help of the scientific method. (Some might say *assumes* is the wrong word here. They would argue that these are all things we have discovered, or they are assumptions that are now adopted in all scientific disciplines.) Bunge also points out that, "contrary to widespread belief, the scientific method does not exclude speculation: it only disciplines imagination." The end product of scientific research is "expected to match reality—i.e., to be true to some degree. Pseudoscientists are not to be blamed for exerting their imaginations, but rather for letting them run loose."[19]

He points out that the

> scientific method presupposes that everything can in principle be debated and that every scientific idea must be logically valid. . . . Furthermore, the scientific method cannot be practiced consistently in a moral vacuum. Indeed, it involves the ethos of basic science, which [philosopher] Robert K. Merton . . . characterized as universalism, disinterestedness, organized skepticism, and . . . the sharing of methods and findings.[20]

Finally, Bunge points to "four more distinguishing features of any authentic science: changeability, compatibility with the bulk of antecedent knowledge, partial intersection with at least one other science, and control by the scientific community."[21]

Some of these characteristics appear in previous chapters, so I only mention a few of his thoughts about them here. A scientific idea has to have some compatibility with prior knowledge "not only to weed out groundless speculation but also to understand the new ideas as well as to evaluate it." Bunge points out that indeed some of the worth of a hypothesis or new experimental design is gauged by the extent to which it fits in with reasonably well-established knowledge: "For example, telekinesis is called into question by the fact that it violates the principle of conservation of energy."[22]

Another, the role of the scientific community, Bunge explains this way: "Investigators do not work in a social vacuum." They experience the thoughts, both positive and cautionary, of their fellow workers; they share problems and findings; and they ask for criticisms; "they get both solicited and unsolicited opinions." This "interplay of cooperation and competition" generates new ideas and controls the spread of results: "It makes scientific research a self-doubting, self-correcting, and self-perpetuating enterprise." This gives science an "ability and willingness to detect error and correct it."[23]

These are features of genuine science. What do we see instead with pseudoscience? Bunge again offers a guide:

> In contrast, a *pseudoscientific* treatment . . . violates at least one of the above conditions, while at the same time calling itself scientific.
>
> It may be inconsistent, or it may involve unclear ideas. It may assume the reality of imaginary facts, such as alien abduction or telekinesis. . . . It may postulate that the facts in question are immaterial, inscrutable, or both. It may fail to be based on previous scientific findings. It may perform deeply flawed empirical operations, such as ink-blot tests, or it may fail to include control groups. It may fake test results, or it may dispense with empirical tests altogether.[24]

Still other characteristics Bunge laments are the way the pseudosciences don't welcome criticism but instead attempt to fix belief. Their basic aim, he says, is not to search for truth but to persuade one to believe in a preconceived idea. He concludes his wide-ranging essay,

> Pseudoscience is just as philosophically loaded as science. Only, the philosophy inherent in either is perpendicular to that ensconced in the other. In particular, the ontology of science is naturalistic (or materialist), whereas that of pseudoscience is idealist. The epistemology of science is realist, whereas that of pseudoscience is not. And the ethics of science is so demanding that it does not tolerate the self-deceptions and frauds that plague pseudoscience.[25]

In 2013 came another major effort to reconsider the demarcation problem—and to defend its importance. It was in the form of a valuable new scholarly book, *Philosophy of Pseudoscience*, edited by philosopher-biologist Massimo Pigliucci (City University of New York) and philosopher Maarten Boudry (Ghent University, Belgium).[26] It presents essays by twenty-six different scholars, most but not all of them philosophers. Lead editor Pigliucci has two doctorates in biology and spent the first part of his career in the United States as a professor of botany and of evolutionary biology (along the way writing several books defending evolution against creationism and sense against nonsense and frequently writing thoughtful columns and articles about the scientific process). He then decided to do an about-face, get a third doctorate (in philosophy), and concentrate on that field, so he is perhaps the ultimate prototypical philosopher of science.

I like Pigliucci and value his work, both because of the unique background he brings to it and because his writings have long exhibited a deep understanding of the processes of science and a balanced perspective on public misunderstandings of it. From 2003 to 2015, he wrote a regular column in the *Skeptical Inquirer* (seventy-two total columns) titled "Thinking about Science." I looked forward to every submission. Luckily, in 2022, he returned with a new column in the *Skeptical Inquirer*, "The Philosopher's Corner," focusing on skepticism from the viewpoint of philosophy. His first topic: "Science Denialism Is a Form of Pseudoscience."[27]

In the 1980s, the American philosopher and epistemologist Larry Laudan wrote a much-discussed paper dismissing the demarcation problem as an ill-conceived and even dangerous pseudoproblem. (A quick aside: *Epistemology*, a term I repeat in this chapter in various forms, for better or worse, is just a grand word for the study of the nature of knowledge, especially its validity and limits.) That paper, "The Demise of the Demarcation Problem," proved highly influential. Laudan says the problem is both "uninteresting" and "intractable." Three decades later Pigliucci and Boudry decided Laudan's dismissal was "premature and misguided" and solicited the essays, their own included, that became the book *Philosophy of Pseudoscience* (2013), to provide a more up-to-date and accurate and realistic view.[28] The lively volume provides, as they say, a constructive discussion about demarcationism among philosophers, sociologists, historians, and professional skeptics. And right at the beginning, they propose a "new philosophical subdiscipline, the Philosophy of Pseudoscience." In so proposing that, they "hope to convince those who have followed in Laudan's footsteps that the term 'pseudoscience' does single out something real that merits our attention."[29]

Studying the nature of science and the differences between science and pseudoscience is crucial, they argue, first because science drives so many revolutionary changes in society and second because pseudoscience is too often wrongly considered a "harmless pastime indulged in by a relatively small number of people with an unusual penchant for mystery worship. This is far from the truth."[30] They point out that creationism, alternative medicines like homeopathy, conspiracy theories about AIDS, denialism about climate change, dangerous cults like Scientology, and all manner of other beliefs in the paranormal and in conspiracies have done enormous damage to public education in the United States and elsewhere. They have swindled people and caused great emotional distress; AIDS disbelief has literally killed hundreds of people around the world.

Pigliucci and Boudry say, "Pseudoscience can cause so much trouble in part because the public does not appreciate the difference between real science and something that masquerades as science." All sorts of other factors come into play, including distrust of academic authorities, the way pseudoscientists parrot scientists, and (something I add here) wishful thinking to have nature and the world be personally understandable and relevant. They claim, "Pseudoscience thrives because we have not fully come to grips yet with the cognitive, sociological, and epistemological roots of the phenomenon." And they give this strong endorsement of the quest's relevance: "This is why the demarcation problem is not only an exciting intellectual puzzle for philosophers and other scholars, but is one of the things that makes philosophy actually relevant to society."[31]

Thus their book: "For all these reasons, we asked some of the most prominent and original thinkers on science and pseudoscience to contribute to this edited volume."[32] They grouped the contributions into six parts: "What's the Problem with the Demarcation Problem?"; "History and Sociology of Pseudoscience"; "The Borderlands between Science and Pseudoscience"; "Science and the Supernatural"; "True Believers and Their Tactics"; "The Cognitive Roots of Pseudoscience." There is too much in the book to fully summarize here, but I give some of the highlights most relevant to this quest into pseudoscience and the shadows of science.

Pigliucci himself starts it off with his examination of the "premature obituary of the demarcation problem." This gets back to Laudan. He writes, "Though much is right in Laudan's analysis, I disagree with his fundamental take on what the history of the demarcation problem tells us: for him the rational conclusion is that philosophers have failed at the task, probably because the task itself is hopeless. For me, the same history is nice example of how philosophy makes progress." Some of Laudan's critique, Pigliucci says, boils down to the argument that no demarcation criterion proposed so far can provide a set of conditions sufficient to define an activity as scientific. Pigliucci argues that Laudan's point has some truth to it, but its "extent and consequences are grossly exaggerated" by Laudan.[33]

At one point, Pigliucci shifts gears from critiquing Laudan to providing some of his own views about what demarcates science from pseudoscience. He proposes that two "threads" run throughout any meaningful treatment of the difference between science and pseudoscience, as well as among further distinctions within science itself. He labels these threads *theoretical understanding* and *empirical knowledge*. He notes that if we can agree on anything about science, it is that science attempts to provide an empirically based understanding of the world. As a result, a scientific theory must have both empirical support and internal coherence and logic. He qualifies this assertion by stating that he does not believe that "these are the only criteria by which to evaluate the soundness of science, but we need to start somewhere."[34]

He even provides a diagram of where various inquiries fit, with theoretical understanding increasing along the horizontal axis and empirical understanding on the vertical axis. Molecular biology, particle physics, evolutionary biology, and climate science reside in the upper right quadrant of Pigliucci's diagram, strong in both aspects, empirical support and theory. They are well-established sciences. We can then move down vertically, where we find evolutionary psychology and string physics, disciplines and notions he finds are theoretically sound but have decreasing empirical content, all the way down at the lower right to superstring theory. This, he notes, is "based on a very sophisticated mathematical theory that—so far at least—makes no contact at all with (new) empirical evidence."[35]

The upper left of his diagram contains the fields of psychology, economics, and sociology, "fields and notions that are rich in evidence, but for which the theory is incomplete or entirely lacking." This includes, he says, many of the social—sometimes called "soft"—sciences. Nothing very controversial so far. It gets more interesting when you move to the lower left corner of Pigliucci's diagram. Here rest astrology, HIV denialism, and intelligent design (he could have added so many others, but he notes those three are just representative). This, he says, is "where actual pseudoscience resides." He points out that they all lie in an "area of the diagram that is extremely low in both empirical content and . . . theoretical sophistication." Take astrology, for example. It has failed all controlled empirical tests of its validity, and while astrologers often proclaim "'theoretical' foundations for their claims, . . . these quickly turn out to be both internally incoherent and, more damning, entirely detached from or in contradiction with very established notions from a variety of other sciences (particularly physics and astronomy, but also biology)."[36]

Pigliucci ends with a pungent challenge to his fellow philosophers on this issue:

> Laudan . . . concluded his essay by stating that "pseudo-science" and "unscientific" are "just hollow phrases which do only emotive work for us. As such, they are more suited to the rhetoric of politicians and Scottish sociologists of knowledge than to that of empirical researchers." On the contrary, those phrases are rich with meaning and consequences precisely because both science and pseudoscience play important roles in the dealings of modern society. And it is high time that philosophers get their hands dirty and join the fray to make their own distinctive contributions to the all-important—sometimes even vital—distinction between sense and nonsense.[37]

The German philosopher Martin Mahner laments the lack of interest in the demarcation problem by many of his colleagues but says some things are accepted. (I met Martin a few years back when I visited the German offices of our Center for Inquiry in Roßdorf. He's an amiable and likable scholar.) In *Philosophy of Pseudoscience*, he argues, "Despite the lack of generally accepted demarcation criteria, we find remarkable agreement among virtually all philosophers and scientists that fields like astrology, creationism, homeopathy, dowsing, psychokinesis, faith healing, clairvoyance, or ufology are either pseudosciences or at least lack the epistemic warrant to be taken seriously."[38]

He takes us through an excursion of "Problems of Demarcation" (among them, there are demarcation issues not just between science and pseudoscience but also between science and nonscience and between good and bad science); "Is Demarcation Really Dead?" (Laudan thought so, but no); "Why Demarcation Is Desirable"; and How Demarcation Is Possible." As for the latter, he invokes a cluster approach, not unlike that of Bunge (whose work he references), in establishing a checklist of demarcations, one that is as comprehensive as possible—a "whole battery of science indicators."[39]

He gives his own list similar to those enumerated by Bunge, including:

- Does [the field] accept the canons of valid and rational reasoning?

- Does it admit fallibilism or endorse dogmatism?
- How important are testability and criticism?
- Can the reliability of its methods or techniques be independently tested?
- Are the theories fruitful?
- Are the data reproducible?
- Is the corpus of knowledge of the given field up to date and well confirmed, or is it obsolete, if not anachronistic? [I think of homeopathy here, among others.] Is it growing or stagnating?

He lists fourteen such criteria and acknowledges that the list could be vastly extended. At one point, he parenthetically guesses that there may be thirty to fifty criteria. This "sheer amount of possible indicators," he admits, shows that it is "unlikely that each of them is fulfilled in every case of demarcation," and they are not all equally important.[40]

He advocates this variable "cluster approach to demarcation . . . based on a comprehensive checklist of science/pseudo science indicators" and a realization that they should be "different from field to field." He concludes, "We must indeed say goodbye to the idea that a small set of demarcation criteria applies to all fields of knowledge, allowing us to clearly partition them into scientific and nonscientific ones. The actual situation is more complicated than that, but it is not hopeless either."[41]

Mahner takes note of another of the book's contributors, Sven Ove Hansson of the Royal Institute of Technology in Stockholm, Sweden. Specifically, he refers to Hansson's definition of *pseudoscience*. It is not dissimilar to the definitions I offer in chapter 3. Hansson states,

> A statement is pseudoscientific if and only if it satisfies three criteria:
>
> 1. It pertains to an issue within the domain of science. . . .
> 2. It suffers from such a severe lack of reliability that it cannot at all be trusted. . . .
> 3. It is part of a doctrine whose major proponents try to create the impression that it represents the most reliable knowledge on its subject matter.[42]

This third aspect is important. And Mahner points out that this "epistemic warrant" can only be a shorthand term for a more extensive list of criteria, such as he himself proposes.[43]

Hansson begins his chapter in amusing style: "For a scientist, distinguishing between science and pseudoscience is much like riding a bicycle. Most people can ride a bicycle, but only a few can account for how they do it. Somehow we are able to keep balance, and we all seem to do it in about the same way, but how do we do it?" As Mahner also acknowledges and as I point out earlier, scientists do this naturally and subconsciously. Writes Hansson, "Scientists have no difficulty in distinguishing between science and pseudoscience. We all know that astronomy is science and astrology not, that evolution theory is science and creationism not, and so on." He notes, "Scientists can draw the line between science and pseudoscience, and with few exceptions they draw the line in the same place. But ask by what general principles they do it. Many of them find it hard to answer that question."[44]

Bristol University philosopher of science James Ladyman, in his chapter, argues that the "concept of pseudoscience is distinct from that of nonscience, bad science, and science fraud" and also that the "concept of pseudoscience is a useful and important one in need of theoretical elaboration."[45] A third major point he makes is a bit more novel. He refers to American philosopher and Princeton emeritus professor Harry Frankfurt's remark in his 2005 book *On Bullshit*: "One of the most salient features of our culture is that there is so much bullshit."[46]

(So true—so much so that an English philosopher colleague, Stephen Law, titled his 2011 book on how to use critical thinking to ferret out nonsense *Believing Bullshit: How Not to Get Sucked into an Intellectual Black Hole*. Since then, two University of Washington academics, Carl T. Bergstrom, a biologist, and Jevin D. West, a communications researcher, titled their 2020 book *Calling Bullshit: The Art of Skepticism in a Data-Driven World*. They have since been calling on the scientific community to more systematically confront misinformation in and about science.[47])

Ladyman says progress can be made in thinking about pseudoscience by learning from Frankfurt's account of bullshit: "Bullshitting, according to Frankfurt, is very different from lying. Pseudoscience is similarly different from science fraud. The pseudoscientist, like the bullshitter, is less in touch with the truth and less concerned with it than either the fraudster or the liar."[48] It reminds me, as a onetime newspaper journalist, of a good old-fashioned newspaper editor's most important tool: the bullshit detector. I argue that successful people in many fields (journalists, lawyers, law enforcement officials, scientific skeptics, consumers) must have this detector. If they don't, then they should. And it may be just as difficult to describe how it works as it is to describe how we ride a bicycle.

Next, let me introduce you to Noretta Koertge. After getting her bachelor's and master's degrees in chemistry at the University of Illinois, she went to the University of London, Chelsea College, for a PhD in the philosophy of science and worked with Karl Popper. She has focused on scientific methodology, the development of science, value issues in science, the philosophy of social science, and critiques of postmodernism and of feminist issues in science. She is former editor of the journal *Philosophy of Science*. She has also written a couple novels.

She seems to have a different perspective from some of the philosophers I discuss so far, and that results in one of the more interesting chapters in *Philosophy of Pseudoscience*. She tantalizingly titles it "Belief Buddies versus Critical Communities" (more on that shortly).[49] *Philosophy of Pseudoscience* has many other interesting contributions, but this is the last essay from it that I focus on here. Now an emeritus professor of philosophy at Indiana State University, Koertge goes beyond philosophy and the demarcation problem in her essay to what I say is sociology, in this case the sociology of communities of pseudoscientists. This is essential because as Bunge and many others point out, one of the important characteristics of science is that it is conducted by a community of scientists, and this community subjects all new ideas and observations to all kinds of informal and formal review, positive and negative. In this way, all new ideas and observations are, at a very early stage, given lots of informed criticism, and this usually results in new and better observations and revised and more sophisticated arguments. This is one of the great strengths of science: Long before anything goes public, lots of knowledgeable people have weighed in on any new idea and helped ensure that it has some merit worth sharing more widely.

So Koertge asks this fascinating question: "What would happen if there were also critical communities supporting *pseudoscientific* inquiry?" And a follow-up question: "Is there something inherent in such enterprises that precludes or makes unlikely such organizational structures?"[50]

She begins by looking at two serious contemporary attempts to institutionalize fringe science. The first is the Society for Scientific Exploration (SSE) and its *Journal of Scientific Exploration*. When this organization and journal were founded in the early 1980s, some hoped they would bring new respectability to unorthodox ideas too way out to be considered mainstream science. They hoped they might even show that some fringe ideas actually deserved science's acceptance. It hasn't quite worked out that way.

Koertge characterizes the program of one of the SSE conferences as "advanc[ing] research in a wide range of areas that would usually be labeled pseudoscience." She is forthright in using the word *pseudoscience*. She looks at summaries of talks given at the conference and finds those on psychic phenomena that might well be useful contributions. She finds nothing of value in a talk that asserts that "psychic reception is more accurate when the constellation Virgo is overhead." She finds that almost every paper references "terms common in various pseudosciences, such as 'remote viewing,' 'reincarnation,' 'precognition,' and 'subtle energy.' Predictably, on the third day there are a couple of papers complaining about dogmatism in science."[51]

The second example of institutionalized fringe science she examines is the *Journal of Condensed Matter Nuclear Science*. At first, this may sound like a regular technical journal, but it makes frequent reference to "low-energy nuclear reactions," which is a sanitized term for *cold fusion*. In fact, its goal is essentially to legitimize cold fusion (under a new name), which as most will recall got off to a terribly rocky beginning in 1989 when two orthodox chemists made dramatic claims about achieving "tabletop" fusion in a small chemistry laboratory. All the while, nuclear physicists at national laboratories were struggling to achieve fusion with giant particle-beam accelerators, laser-beam accelerators, or magnetic-confinement fusion projects requiring enormous temperatures and pressures, achieved only in huge national and international science projects of large scale and cost. The claims quickly fell apart, and the two scientists and their scientific claims disappeared into relative ignominy. But it was all started by researchers with traditional scientific credentials. In contrast, Koertge finds that topics discussed in the *Journal of Scientific Exploration* "have always been viewed with suspicion and their proponents in some cases have little relevant scholarly training."[52]

She looks briefly at two more communities of fringe scientists, the Discovery Institute, which promotes intelligent design creationism, and the International Center for Reiki Training. She finds neither promising. All these and other groups she examines have the drawback that they are composed essentially of what she calls "belief buddies." These groups form a community of like-minded believers who "feel that their views are often neglected or stigmatized. As a result, these belief buddies consciously attempt to affirm contributions that further their agenda; dissent is discouraged lest it lead to a splintering of the group." These kind of belief buddies don't welcome criticism. They are there to "reassure beleaguered constituents," not give constructive suggestions for improving their approaches and arguments.[53]

These groups' reluctance to encourage criticism is a great weakness; the "result is that some pseudosciences are plagued with dubious reports from credulous amateurs or even charlatans looking for attention. We have a romantic image of the lone scientist working in isolation" and against all odds overturning previous ideas. But science just doesn't work that way. Not often anyway. Koertge concludes, "If our would-be genius does make a seemingly brilliant discovery, it is not enough to call a news conference or promote it on the web. Rather, it must survive the scrutiny and proposed amendments of the relevant critical scientific community." She then calls on her fellow philosophers to work with scientists in maintaining science's high values "and also [call] attention to the inadequacies of projects that we correctly label as pseudoscience."[54]

I end with a comment by another prominent philosopher of science, Susan Haack. She is distinguished professor in the humanities, professor of philosophy, and professor of law at the University of Miami School of Law. She writes often about how philosophy today is not in good shape and how crucial it is for philosophy to engage with real subjects that matter. When it comes to science, she has made many contributions and emphasizes that there is no single, distinctive, timeless scientific method but many approaches, modes, and procedures.[55] In her 2013 book *Putting Philosophy to Work*, she writes about, among other things, the untidy process of groping for the truth. In doing so, she provides an elegant passage that highlights some of the things I talk about in this chapter, including the importance in science of communities of scientists. She does not write in the context of pseudoscience but emphasizes some of the processes of science that help avoid error and reach truth. And that's why I find her words so pertinent here:

> As human cognitive enterprises go, the natural sciences have been remarkably successful; not because they use a uniquely rational method of inquiry, unavailable to other inquirers, but in part because of the many and various "helps" they have devised to overcome natural human limitations. Instruments of observation extend sensory reach; models and metaphors stretch imaginative powers; techniques of mathematical and statistical modeling enable complex reasoning; and the cooperative and competitive engagement of many people in a great mesh of subcommunities within and across generations not only permits division of labor and pooling of evidence but also—though very fallibly and imperfectly, to be sure—has helped keep most scientists, most of the time, reasonably honest.[56]

As this chapter explains, working scientists might have an intuitive general sense of what constitutes science and what pseudoscience is, but philosophers have always sought to probe the matter more deeply, with finer resolution, and with more formality. The demarcation problem has more than once been prematurely dismissed as either unsolvable or irrelevant, but it is not going away, nor should it. Scientist-philosophers like Bunge and Pigliucci, even though their approaches are different, have revived intellectual interest in the problem. They and a number of colleagues around the world offer penetrating insights into what science is, how it works, and why it is so successful. They acknowledge that no single criterion works in distinguishing science from pseudoscience. But they propose a variety of criteria or clusters of characteristics useful for telling the difference. They also champion the view that philosophers need to continue to give serious attention both to this pesky boundary problem and to the nature and appeal of pseudoscience itself.

CHAPTER NINE
CLIMATE ANTISCIENCE AND DENIAL

One day in the early to mid-1980s, I got a phone call from a boyhood acquaintance. He was now a successful farmer, rancher, and businessman in Colorado. He was considering buying a cattle ranch in southeastern New Mexico near Roswell, and he was concerned about the warming climate. He had heard I might know something about this, so he asked, "How much do you think global warming is going to affect the cattle business down there?"

He was serious. There was not a hint of doubt in his mind that the reports he had heard from scientists about global warming were real. For his business, he had to take them into account. I assume he was a political conservative and a rock-ribbed Republican. Most ranchers in the West are. But global-warming denial was not yet a thing. It hadn't taken hold of a vast segment of America's political culture.

How things have changed. By at least the mid-1990s, a vast campaign to discredit the scientific findings of climate scientists—in fact, to denigrate that entire scientific field and even to personally attack scientists researching climate—was underway. Climate scientists were amassing ever more evidence that the world was warming; that as a result, our entire climate was changing; and that this process would continue and worsen. It would change our world in unpredictable ways if we couldn't figure out how to curtail the continually rising emissions of greenhouse gases.

Yet a strong wave of political opposition to these conclusions had arisen with equal force and vigor, and an antiscience battle of near-epic proportions was in progress. Some have called it the Climate Wars. It is still underway, albeit differently. The influence and intensity of this anti-climate-science campaign took scientists by surprise. It also surprised most people who follow science closely. Scientifically, it made no sense. Climatology was a maturing science. The lines of evidence supporting global warming were multiple and reinforcing. There was really nothing scientifically controversial about the amassing evidence that the planet was warming. It wasn't unexpected. In fact, it had been predicted. It was the result of documented circumstances and known scientific processes. And once global levels of greenhouse gases in the atmosphere were shown to be steadily rising, warming seemed almost inevitable. Now those forecasts were being validated by global measurements of increasing temperatures. The Earth was warming. *That* was the issue. Shouldn't we focus on what to do about *that* problem rather than trying to discredit the scientific results that were calling it to our attention? That's called "shooting the messenger."

What a naïve view this was. (In hindsight, we can admit that.) We are all prone to such naïveté. Scientists employ evidence, and they try to use reason and rationality. They keep being blindsided when people don't willingly accept their logical and evidence-based findings. All of us who try to think in scientific ways are vulnerable to getting whacked on the behind by those whose priorities lie in other directions. It happens repeatedly, over and over and over. You'd think we'd learn.

In chapter 2, I point out that antiscience differs from pseudoscience in a fundamental way: While pseudoscientists usually try to mimic science and bask in the glow of science's general credibility, antiscience actively *opposes* good science, or at least some parts of science. For a variety of ideological reasons, it tries to undermine unwelcome scientific findings and even discredit the scientific processes that lead to them. Often it is backed by well-funded opposition groups who set up sham think tanks and employ influential lobbyists and public relations firms to influence public opinion. And they also often focus specifically on political leaders and decision makers. That moves their antiscience thinking into the policymaking process. Some of our legislative and administrative decisions are tainted with it. As a result, antiscience can be even more pernicious and vicious—and more detrimental to our democratic processes—than out-and-out pseudoscience.

Climate science has been the most active target of the antiscience efforts in recent decades. It may be difficult for us science-minded Americans to admit, but while the findings of climate science are widely accepted in the rest of the world, the anti-climate-science movement is especially strong in the United States. It has entrenched itself into the American political process with uncommon effectiveness. It's discouraging to have to acknowledge this fact. Many Republican congressmen are out-and-out climate deniers, and most of the rest go along with that viewpoint to varying degrees, as they've found it is either necessary politically or that it works for them in getting votes or key committee appointments. It has become a kind of litmus test for party loyalty. How the venerable Republican Party became an antiscience party (at least on climate change and a few other topics, like evolution) is a lamentable question to have to ask, but that is not my focus.[1]

No political party or persuasion has a unique hold on reason and high-minded virtue. Many liberals, for instance, distrust the science that tells us GMOs are safe and important to sustaining the Earth's population or that nuclear power is an energy-efficient alternative to fossil fuels and adds no carbon dioxide into the atmosphere. Some who oppose life-saving vaccines are also overly open to a lot of pseudoscientific, New Agey "woo-woo."

As a veteran science journalist, I have reported on climate science all my life. I have always loved the sky, clouds, and weather. At *Science News*, I first served as the Earth sciences editor, which encompassed covering all new science about the Earth, oceans, and atmosphere and how they all interact with each other. I loved covering the Earth and geophysical sciences, and weather and climate are one key aspect of that. I continued to write about both after becoming *Science News*'s chief editor and even those years afterward, when I had a five-year stint as a contributing editor. My first book, *The Violent Face of Nature*, deals with severe weather and the natural disasters that result when human populations find themselves in the crossfire.[2] To this day, I am still a weather nerd. I duly report any daily precipitation captured by my professional rain gauge (to a hundredth of an inch) to an interregional rain-reporting network.

Before continuing, I want to state the differences between weather and climate. Weather is the totality of atmospheric conditions at any place and time, the instantaneous state of the atmosphere that especially affects living beings. Climate is the sum of the weather experienced over the years. It is, in essence, the long-term average of the weather.[3] I mention this for a particular reason.

Weather forecasts we see every night on our local news and climate projections are very different things. And for a while, a few years back, it seemed that some television weather meteorologists were having trouble accepting climate science. I think this was partly due to their focus on *weather* forecasting. Due to dramatic improvements in observing technologies over the past few decades (and in the science of forecasting), accurate weather forecasts for two, three, and sometimes even ten days are now routine. But doing that same kind of forecast for years or centuries in the future is clearly impossible. I think that is what some of the forecasters were thinking of and finding implausible.

But of course, that's not what climate scientists do. Their future projections are of long-term trends based on long-term statistical averages of temperature, and they make no pretense to the kind of near-term local and regional specificity of actual *weather* that weather forecasters do. This situation has vastly improved now, and almost all weather forecasters now accept that climate change is real and happening. In fact, a number of them have become science educators on climate. And they can be very effective in this role, as they generally have their viewers' trust.

Two Iconic Scientists: Roger Revelle and Stephen H. Schneider

Two scientists strongly influenced me early on in my appreciation of climate science: Roger Revelle and Stephen H. Schneider. Both had broad knowledge, a global vision, and a willingness and ability to speak out about what the science was showing, but they were also very different.

Revelle was much older and long a respected statesman of science. I would see and hear him speak at National Academy of Sciences meetings and conferences of the American Association for the Advancement of Science (AAAS), where he spoke nearly every year and served a term as president. He was the consummate institutional man, serving on all kinds of national boards and committees setting the directions of science. He helped establish both the groundbreaking International Geophysical Year in 1957–1958 and the University of California, San Diego, campus (where Revelle College is named for him). A big man, usually in a suit, he moved slowly, but when he spoke, people stood up and listened. An oceanographer by training (he formally switched to public policy later), he was always an articulate messenger about how the oceans and atmosphere interacted, how important it was to understand all these processes, and how the climate could change.

In 1957, Revelle and Hans Suess of the Scripps Institution of Oceanography published these prophetic words about how burning fossil fuels was likely to change global climate: "Human beings are now carrying out a large scale geophysical experiment of a kind that could not have happened in the past nor be reproduced in the future. Within a few centuries we are returning to the atmosphere and oceans the concentrated organic carbon stored in sedimentary rocks over hundreds of millions of years."[4]

The paper goes on to suggest that the result might be a "greenhouse effect" that would eventually cause global warming. In 1982, Revelle published an influential *Scientific American* article on the carbon dioxide issue.[5]

He remained an active champion of these issues until late in life. My son, Chris, a science graduate of the University of California at San Diego, told me recently that when he was a graduate student at San Diego State University, he arranged for Revelle to speak on campus. It was one of Revelle's last public appearances. What did he talk about? Global change.

In 1990, Revelle went to the White House to receive the National Medal of Science from President George Bush. He told a reporter, "I got it for being the grandfather of the greenhouse effect."[6] Revelle died in 1991 at the age of eighty-two.

When I first knew him, Schneider was, in contrast, young, slender, telegenic (with curly dark hair). He always had a way with words. He had much of the appeal and charisma of Carl Sagan, and he often appeared on television, passionately advocating for climate science and other science-related issues. Back then, he directed climate studies at the National Center for Atmospheric Research in Boulder. Later he moved to Stanford. He was comfortable working closely with the news media and science journalists (that's how I met and got to know him), and I think we considered him almost one of us. His first book, *The Genesis Strategy*, was about climate and global survival. In 1984, his massive book *The Coevolution of Climate and Life* devoted more than a third of its bulk to climate change. By 1989, he had an entire book on the subject, *Global Warming: Are We Entering the Greenhouse Century?*[7]

Strong proponents who put public faces on the issues of climate science (as climatologists like James Hansen and Michael Mann would also later do), Revelle and Schneider became active targets of the anti-climate-science movement. Schneider seemed almost to relish the battle. He was very good at it. He testified to congressional committees, spoke on Sunday morning network television news shows, and was tireless in taking the fight into the public arena. It is a quality very few scientists have, and understandably so. The title of Schneider's last book was about that fight, *Science Is a Contact Sport*. It not only defends climate science against its opponents but also gives vivid, blow-by-blow accounts of the attacks, infightings, and delays that have slowed any reasonable response.[8]

One more relevant insight from Schneider: In 2009, I heard him speak at the AAAS meeting in Chicago on media coverage of climate change. One point he made struck me as so important that I asked him to let us publish it in the *Skeptical Inquirer*. He revised it and sent it to me immediately. It addresses that aggravating tendency for climate deniers to try to call themselves climate skeptics. We science-minded skeptics feel *skeptic* is a noble term. It must be reserved for those who provisionally question unsupported claims but go with the evidence and accept good science once fully supported and established.

Here is some of what he wrote:

> All good scientists are skeptical. I changed my mind from cooling to warming in 1974 when the preponderance of evidence shifted—and is now well established. . . . Real skeptics still accept a preponderance of carefully examined evidence even when some elements of a complex systems problem remain unresolved, and they do not pretend that when there are loose ends some well-established preponderances don't exist—that is beyond skepticism to denial—or often political convenience. So a skeptic questions everything but accepts what the preponderance of evidence is, and a denier falsely claims that until all aspects are resolved we know nothing and should do nothing—often motivated by the latter. If you deny a clear preponderance of evidence, you have crossed the line from legitimate skeptic to ideological denier.[9]

Steve Schneider died suddenly of an apparent heart attack while traveling in London in July 2010. He was only sixty-five. It was a terrible loss to the climate science community and to the public understanding of science generally.

At any rate, I became familiar with climate researchers long ago. I realized their work was important and timely. I respected them and shared their values and concerns. A subset of them are paleoclimatologists, who have gallantly deciphered the record of past climates. To do so, they have used all sorts of clever tools, including isotopic analysis of tree rings, ice cores, ancient coral reefs, and marine skeletons buried in seafloor sediments. These can tell us not only the record of changing past temperatures but also—even more astonishing—the varying volume of glacial ice on land. Geologists then calibrate all this with their own studies of past changes in sea levels, and with all the other evidence from other fields, you soon get a mutually reinforcing and relatively consistent picture of Earth-temperature history.

I have also read histories of how changing climates in the past have severely affected human history. Two I especially recommend are Robert Clairborne's *Climate, Man, and History* and Brian Fagan's *The Long Summer: How Climate Changed Civilization.* You might also want to read Fagan's *The Little Ice Age.*[10]

The Dance of the Ice Ages

Through their studies, Earth scientists and climatologists have detected, identified, and explained various key climate cycles. One of these is, to my mind, an amazing scientific story. It is how mathematician Milutin Milankovitch (1879–1958) put on firm quantitative footing the astronomic explanation of the major ice ages that have come and gone throughout much of Earth's history. Oceanographer John Imbrie and Katherine Palmer Imbrie tell this story in their book *The Ice Ages: Solving the Mystery*, and I relate it in a chapter titled "The Dance of the Orbits" in a book I wrote about the Sun and its interactions with Earth.[11] I want briefly to recall it here.

Milankovitch, born in what was then part of the Austro-Hungarian Empire and is today Croatia, got his doctorate in 1904 from the Vienna Institute of Technology. As a young mathematician in Belgrade, Milankovitch found himself inspired: "I feel attracted by infinity. . . . I want to grasp the entire universe and spread light into its farthest corners." The "cosmic problem" he decided to tackle was a mathematical theory that would allow him to calculate the past and present temperatures and climates of the Earth.

Astronomers already knew that there are cyclic changes in the Earth's orientation to the Sun. For instance, Earth's North Pole doesn't always point to the same spot in the sky (now very near Polaris, the North Star). The Earth slowly wobbles like a top, and it completes one such wobble every 25,800 years. (In 120 BCE, Hipparchus compared where the pole was pointing then to where astronomers had said it had pointed 150 years earlier and took note of this wobble. I find that observation extraordinary, as well.) Our planet's axis describes an orbit of 47 degrees over the entire 25,800-year period. So that's one cycle. This axial precession is slightly countered by the fact that Earth's elliptical orbit is itself, even more slowly, rotating in space in the opposite direction. That motion cancels out some of the effects of axial precession and results in a combined cycle lasting about 22,000 years, known as the precession of the equinoxes.

Another cycle is one we've known about since 1843, when the great French astronomer Urbain Jean Joseph Leverrier demonstrated it: The shape of Earth's orbit—how much it is out of round, its eccentricity—varies slightly over a roughly 100,000-year cycle. Sometimes our orbit's eccentricity is nearly 6 percent (a bit elliptical), other times, near 0 percent (almost round). In our era, it is 1.7 percent and decreasing. When Earth's orbit becomes more eccentric, this increases the magnitude of seasonal changes. This 100,000-year cycle turns out to have a strong influence on cycles of the ice ages, where there is a strong 100,000-year pattern.

Leverrier was also responsible for the first insight that led to quantifying a third cycle: the slowly changing tilt of Earth's axis. Earth's axis is now tilted 23.5 degrees from vertical (from being perpendicular to the plane of Earth's orbit). This is what gives us our seasons. In summer, Earth's Northern Hemisphere tilts toward the Sun, giving longer days and more direct sunlight. In winter, the Northern Hemisphere tilts away from the Sun, giving shorter days and less direct sunlight. (The reverse is true, of course, for the Southern Hemisphere.) Leverrier calculated that this amount of axial tilt actually varies over time between 22 degrees and 24.5 degrees. He did not calculate the timing, but we now know that the angle of tilt of our axis varies over a cycle of 41,000 years.

Milankovitch took it upon himself to calculate the various effects on the Earth of these three major cycles: precession, eccentricity, and tilt. His work was interrupted first by the Balkans War in 1912 and then World War I (where he was briefly captured as a prisoner of war by the Austro-Hungarian army), but he then settled into the Hungarian Academy of Sciences in Budapest and continued his work. He published a book in Serbian in 1920 that few read and then finally, in 1930, the book that wrapped it all up. He had solved his cosmic problem.

Over the next decade or so, these Milankovitch cycles began to fade into oblivion as just a curious matter. But they were eventually revived and strongly confirmed (finally, in 1976) by the work of a whole host of Earth and ocean scientists doing studies of actual Earth temperatures and of the advances and recessions of the ice sheets. This work—first by such people as Wallace S. Broecker of Columbia University and Cesare Emiliani of the University of Chicago, then by Cambridge University geophysicist Nicholas J. Shackleton, Brown University geologist-oceanographer John Imbrie, and Columbia University microfossil expert James D. Hays—gave the astronomic theory of ice ages its first sophisticated test. They got their temperature data from two sediment cores from the bottom of the southern Indian Ocean that together gave a record of temperature variations for the past 450,000 years. The overall agreement between the newfound climate record and the calculated astronomical curves was remarkable. Each of the cycles from the cores matched the predicted cycles to within 5 percent. Hays, Imbrie, and Shackleton published their paper "Variations in the Earth's Orbit: Pacemaker of the Ice Ages" in the December 10, 1976, issue of the respected journal *Science.*[12] Of these three major cycles—100,000, 41,000, and 23,000 years—the 100,000-year distance-effect oscillation seems to have been dominating the variations in sunshine and seasonality over the past half-million or so years. Ice cores and marine fossils in sedimentary cores each separately clearly show temperature and ice advance and retreat cycles of 100,000 years.[13]

So the Milankovitch cycles seem largely to be the pacemaker of the ice ages over the past two or three million years. But what about before that, when we didn't have such ice ages? These cyclical orbital variations would have been in effect earlier, as well. What else could have been different? Well, it is very likely the positions of the continents. The scientific revolution of plate tectonics (which I also had the privilege of covering) showed that the continents have moved about the Earth's surface on the backs of enormous underlying crustal plates. Over tens of millions and hundreds of millions of years, their shapes, sizes, orientation, and location on the Earth have changed dramatically. It turns out that in our current era, most of the continental surface area is in the Northern Hemisphere, and three major continents (North America, Europe, and Asia, plus Greenland, now still mostly covered with an ice sheet, although its ice is rapidly melting) crowd up against the Arctic Ocean. For a variety of reasons, including the ample snowfall that can happen, this situation is very conducive to having periodic ice ages on these northern continents. Before three million years ago, the continents were not quite so crowded in the Arctic Ocean, and before that still they were all much farther south, which may have lessened the likelihood to initiate cycles of ice ages.

It's Not Weather, and It's Not Just Cycles

I wanted to give you this little flavor of just one specific example of the science that climate scientists—broadly defined—do and of the discoveries and insights they have come up with. As I mentioned earlier, to contrast science with either pseudoscience or antiscience—as I do in this book—it is good to get familiar with the science and how scientists achieve their results. Multiply the story of the discovery and confirmation of the Milankovitch cycles by thousands of other such examples of research findings in recent times, and you can see why climate science is such an intellectually important and fruitful scientific field.

Notice also that *climate science* is a collective term for all fields of science that contribute to understanding our climate and its changes: physics, chemistry, astronomy, mathematics, oceanography, the atmospheric sciences, meteorology, geology, engineering, and many more. Those trained in atmospheric sciences who specialize in climate obviously are climate scientists. But so, too, are the scientists from countless other fields who bring their specific expertise to bear on climate issues. They all publish peer-reviewed scientific research on climate. So when you hear of attacks on climate scientists as though that's just a narrow new specialty that perhaps we needn't be concerned about, know that it really is science overall that is being targeted.

Also, it's the matter of cycles I want to raise. Those who dispute the evidence of global warming are fond of saying, "It's all just cycles." Their idea is that, yes, the Earth might be warming just now, but it's all part of a natural cycle that will soon reverse itself. Two points about that: Yes, there are cycles, and that's what climate scientists have been carefully studying and documenting for decades now. As with the important long-term Milankovitch astronomical cycles, they are well aware of them. They understand them. They are the ones who discovered them.

Here are examples of some short-term cycles: The most important ocean-atmosphere cycle is the El Niño–Southern Oscillation (ENSO), with its two modes, El Niño and La Niña. Two others play out over the North and South Poles and affect all the surrounding land and sea. The Northern Annular Mode (NAM) and the Southern Annular Mode (SAM) are oscillations in the strength of the upper-level winds that circle the globe at these high latitudes. In the North Atlantic, NAM becomes the North Atlantic Oscillation (NAO)—pressure variations between Iceland and the Azores. Ocean cycles include the Pacific Decadal Oscillation (PDO) and the Atlantic Multidecadal Oscillation (AMO).[14]

Climate scientists are well aware of all these cycles and more. What is new and so troubling is a more or less steadily upward progression of temperatures that is unprecedented in human history. This upward trend is superimposed over all the various short-term cycles of smaller ups and downs and threatens to overwhelm them. That is what is happening. So for opponents to argue that what is at work are merely cycles and to imply that somehow climate scientists who warn us about a warming planet are unaware of or ignoring these cycles is the grossest logical contradiction. Do they really think scientists would be warning us so strongly about the recent upsurge of warming if they didn't understand that this is something over and beyond the normal, routine cycles of climate? That is nonsensical.

A few years ago, I heard a talk by the noted climate scientist Henry Pollack (a University of Michigan professor of geophysics and author of *A World without Ice*). He got a question about this "natural variability." He was frustrated when he responded, "Natural cycles? The natural factors are still at play. But the human player is new. It is outstripping the natural variability."[15]

Why So Sure? The Evidence Amasses

Here is why scientists are so certain this upward trend will continue. One is theoretical; one, observational.

1. The Theory Showing That Rising Carbon Dioxide Levels Should Lead to Rising Atmospheric Temperatures

In the 1850s, the English physicist John Tyndall (1820–1893) measured the ability of various gases, including water vapor and carbon dioxide, to absorb infrared radiation (heat). This kind of greenhouse effect had been surmised before, but Tyndall was the first to prove it. (Tyndall also advocated for clear separation of science and religion. In addition, he became a strong supporter of Charles Darwin and his then new theory of evolution by natural selection. He knew and was close friends with Thomas H. Huxley, Darwin's bulldog, and as president of the British Association for the Advancement of Science, he spoke out on behalf of Darwin's ideas repeatedly.)

Next came the Swedish physicist-chemist Svante Arrhenius (1859–1927), a founder of the field of physical chemistry. Arrhenius was a mathematical prodigy. He wanted to find out if greenhouse gases could help explain the temperature variations of glacial and interglacial periods in past Earth history, so he calculated how much infrared, or heat, radiation is captured by CO_2 and water vapor in Earth's atmosphere. His equation found that as the quantity of carbon dioxide increases geometrically, the atmospheric temperature increases arithmetically. His elegant formula for that rule from 1896 is still in use today. Arrhenius thus was the first person to use basic principles of physical chemistry to calculate the extent to which increases in atmospheric carbon dioxide would increase surface temperatures through the greenhouse effect. He found that these effects should be large enough to create global warming. This theory of his has since been extensively tested, and it has held up well. (In 1903, Arrhenius received the Nobel Prize in physics, but it was for other findings of his in physical chemistry, not this work.) So for more than a 120 years, we have had a strongly supported theory for how rising atmospheric carbon dioxide should lead to rising atmospheric temperatures.

I think the power of this effect is very difficult for most people to appreciate. How can molecules of an invisible gas we cannot even see and that we wouldn't even know about without measuring have such an effect? I can understand that thinking. That is a very human reaction, but the fact is, the effect is real and valid. It may be nonintuitive, but science is all about trying to find out what is real, and that often *does* go against our presuppositions. Otherwise, we wouldn't need science; we could just go by our own experience on all things. The term *greenhouse effect* (not fully accurate in itself but nevertheless useful) does, I think, help most of us to visualize how this can happen. Without it, I think public understanding of the situation would be even more difficult.

It might help to realize that the amount of CO_2 we've pumped into the atmosphere is massive. One calculation done in 2007, when the concentration was 383 parts per million, is that there are more than 3 trillion tons of carbon dioxide in the atmosphere (specifically, 2.996×10^{12} tonnes; a tonne is a metric ton, 1,000 kilograms, or 2,204.6 US pounds).[16] Today the situation is far worse. Carbon dioxide measured at the National Oceanic and Atmospheric Administration's (NOAA) Mauna Loa Atmospheric Baseline Observatory in Hawaii peaked at 421 parts per million in May 2022, pushing the atmosphere further into territory not seen for millions of years.[17] If you think about how lightweight carbon dioxide gas is, you can realize how truly enormous this amount of greenhouse gas must be. That amount has only gone up since then. CO_2 levels in the atmosphere are now 50 percent higher than their preindustrial levels.[18]

It is all difficult to visualize. Carbon dioxide is the product of combustion. If it were visible, we would see it ourselves, spewing out the backs of our vehicles and out of our furnaces and power plants like the visible dark soot and particulates we used to see coming out of chimneys and exhausts. The 2014 television documentary *Cosmos: A Spacetime Odyssey* (with Ann Druyan as main writer and producer and Neil deGrasse Tyson as host) has an episode where they used animation to portray the carbon dioxide we emit.[19] Clouds of yellow gas spewed out everywhere. It was a powerful set of images.

2. The Continuing Upward Rise of the Amount of Carbon Dioxide in the Atmosphere

It is now at record levels for historic times. This realization is again the result of an impressive scientific activity, the long-term record of measurements of atmospheric CO_2 carried out since 1958 atop Mauna Loa Observatory on the mid-Pacific island of Hawaii. Chemist Charles David Keeling convinced the Scripps Institution of Oceanography to begin this measurement program as part of the International Geophysical Year (IGY, 1957–1958). The IGY was an epic eighteen-month multination research effort that greatly advanced the understanding of our planet and its systems. (I was very young then, but the IGY helped inspire my love of the geophysical sciences.)

Keeling realized that carbon dioxide in the atmosphere would be well mixed and that a perfect place for measuring it would be the middle of the Pacific Ocean, well away from local influences. He set in motion a series of precision measurements that continue to this day. The measurements first clearly show a small, sawtooth-shaped seasonal signal due to vegetative growth in the Northern Hemisphere, which is testimony to its precision. But from virtually the beginning, the overall trend of this "Keeling curve" has been steadily upward. Carbon dioxide levels rose from about 315 parts per million in 1958 to around 370 parts per million in 2000. In 2013, the levels surpassed 400 parts per million for the first time in human history. By the end of 2017, they had reached 405 parts per million. By September 2018, they had reached 406.99 parts per million. And again, by August 2022, they were at 417.19 parts per million, 50 percent higher than preindustrial levels.[20]

In case you might think this rise doesn't seem so large, well, think again. A telling graphic from NOAA displays the record of carbon dioxide levels over the past 400,000 years up to now. Over that time, they rose and fell several times between 180 and 280 parts per million but never were higher than 300 parts per million—until modern times. On this timescale of hundreds of thousands of years, the recent rise is nearly straight up. Levels began rising dramatically in industrial times, and by 1950, they already exceeded levels unseen in the last half-million years.[21]

In fact, NOAA has extended that past record of atmospheric carbon dioxide levels back another 400,00 years, covering all the ice ages and warm periods. It shows that carbon dioxide levels are higher now than at any point in at least the past *800,000* years. The last time CO_2 levels were this high was three million years ago. And guess what? Sea levels were then fifteen to twenty-five meters (fifty to eighty feet) higher than today![22]

The Evidence Is Overwhelming

Taken together these two ideas—science showing that greenhouse gases in the atmosphere should result in rising temperatures and a reliable long-term observational record of steadily rising atmospheric carbon dioxide over the past six decades—gave scientists firm confidence that temperatures must be going up. And that of course is what measurements show is happening. NOAA and NASA have very detailed daily temperature records going back 135 years, so we can compare average regional or global temperatures today to the long-term average in those records.

Here is one key conclusion: *The world has not had a cooler-than-average year since 1976. In fact, it hasn't had a cooler-than-average* month *since 1985!* Think about that. Think about things that fluctuate in a normal way, and consider how extraordinarily unusual it is not to have a single month cooler than the average in a third of a century! It would be like flipping a coin and having it come up heads four hundred times in a row! You'd realize something other than chance was going on. Every year so far in the twenty-first century has been at least 0.75° Fahrenheit (0.4° Celsius) warmer than the twentieth-century average. Every year so far in the twenty-first century is among the twenty-five hottest years on record, according to NOAA's figures.

As for the Arctic and the polar sea ice, well, among other things, it is an important indicator of climate change. Climatologist and ice expert Pollack put it pungently in the talk I heard him give: "Ice is a natural indicator. It's one of nature's own thermometers. It is neutral. It asks no questions. It listens to no arguments. It has no political baggage. It just melts. It's an apolitical indicator."[23] NASA reports that sea ice is declining at a rate of 13.2 percent a decade. The sea ice continues to decline markedly in the Arctic decade after decade, and its thickness is now less than 50 percent of what it was forty years ago, according to scientist Ted Scambos with the National Snow and Ice Data Center. The fabled Northwest Passage through the Arctic was open for a while for the first time in both 2008 and 2009. "We have entered into a remarkable new domain," Pollack says. "No human has ever seen an ice-free Arctic Ocean." He says what's happening in the Arctic is the "biggest change in appearance of the planet that humans have ever seen." And "it's all happening in one generation."[24]

The 2018 Intergovernmental Panel on Climate Change's (IPCC) special report seemed only to put an official stamp on the matter. This was the most up-to-date consensus view from the world's climate scientists working collaboratively as the IPCC (it won the Nobel Peace Prize in 2007 for its efforts). The report concludes, "Human activities are estimated to have caused approximately 1.0°C [1.8°F] of global warming above pre-industrial levels, with a *likely* range of 0.8°C [1.4°F] to 1.2°C [2.2°F]. Global warming is *likely* to reach 1.5°C [2.7°F] between 2030 and 2052 if it continues to increase at the current high rate."[25] Consider that these are averages of all temperatures at all times of the day and night at all locations over the entire planet.

The IPCC report adds that many land regions were experiencing even greater warming, including two to three times higher in the Arctic. One report had color charts plotting warming by region. Most of the Arctic was depicted with deep purple, indicating warming in winter of 3.0°C (5.4°F) or greater![26] It also found that many other regions of the world have already experienced greater regional warming than this, with 20 to 40 percent of the Earth's population having experienced more than 1.5°C (2.7°F) warming in at least one season. There are multiple effects already.

"We are already seeing the consequences of 1°C [1.8°F] of global warming through more extreme weather, rising sea levels, and diminishing Arctic sea ice," Panmao Zhai, a climate scientist with the Chinese Academy of Meteorological Sciences and cochair of IPCC working group 1, says.[27] Here's how the report puts that: "Temperature rise to date has already resulted in profound alterations to human and natural systems, bringing increases in some types of extreme weather, droughts, floods, sea level rise, and biodiversity loss, and causing unprecedented risks to vulnerable persons and populations."[28]

Keeping the rise to less than 1.5°C (2.7°F) is the IPCC's hopeful, almost-last-resort target, but doing so would require gargantuan efforts. The IPCC found that achieving that goal would require "rapid and far-reaching" transitions in land, energy, industry, buildings, transport, and cities. Human-caused emissions of carbon dioxide would need to fall by 45 percent from their 2010 levels by 2030 and reach "net zero" by 2050. Net zero? That means any remaining emissions would need to be balanced by *removing* carbon dioxide from the air. And that means we would have to have *negative* carbon emissions![29]

When the IPCC's sixth report came out in February 2022, the evidence of immediate and harmful effects of the changing climate was now so evident that its authors dropped some of their predecessors' cautious language:

> Human-induced climate change, including more frequent and intense extreme events, has caused widespread adverse impacts and related losses and damages to nature and people, beyond natural climate variability. Some development and adaptation efforts have reduced vulnerability. Across sectors and regions the most vulnerable people and systems are observed to be disproportionately affected. The rise in weather and climate extremes has led to some irreversible impacts as natural and human systems are pushed beyond their ability to adapt.[30]

This is followed by some even stronger language about the ongoing effects of more frequent and more intense extreme weather events:

> Widespread, pervasive impacts to ecosystems, people, settlements, and infrastructure have resulted from observed increases in the frequency and intensity of climate and weather extremes, including hot extremes on land and in the ocean, heavy precipitation events, drought and fire weather (*high confidence*). Increasingly since [the IPCC's 2013 main report], these observed impacts have been attributed to human-induced climate change particularly through increased frequency and severity of extreme events.[31]

As if to ratify and reemphasize those findings, the warming planet dealt a series of extreme weather disasters to the world's population in the summer of 2022. Europe suffered an intense heat wave, with temperatures in Spain and Portugal reaching well over 37.8°C (100°F) for days on end. The heat killed two thousand people in Portugal and Spain and an additional one thousand people in the United Kingdom. From Africa and China to Cambridge, England, the intense heat caused railroad tracks to bend, forced airport runways to shut down, and melted pavement on roadways.

People in China suffered at least two months of brutal, record-breaking heat. On August 18, the temperature in Chongqing in Sichuan province reached 45°C (113°F), the highest ever recorded in China outside the desert region of Xinjiang. On August 20, the temperature didn't drop below 34.9°C (94.8°F). It was the longest and hottest heat wave in China since national records began in 1961. A weather historian said it was the most severe heat wave recorded anywhere.[32]

In the United States, the heat broke nearly six thousand temperature records across the Pacific Northwest, the southern Plains, the South, and the Northeast. The long-term drought across the West worsened still, and forest fires continued to ravage areas in California, Idaho, Oregon, and Montana. New Mexico, plagued with an exceedingly dry winter and spring, had its largest and most destructive fire on record, consuming 341,000 acres of forest north and east of Santa Fe (the New Mexico fires finally ended when, thankfully, an early and wet monsoon season brought rain from late June into September).

Water levels at Lake Powell and Lake Mead continued to plunge, exposing sunken boats and lost bodies. In other parts of the nation the problem was way *too* much water. St. Louis got two months of rain in six hours. "Thousand-year" rainstorms hit Illinois and eastern Kentucky, where at least thirty-seven people died in the floods. By mid-September 2022, the United Nations said the world was entering "uncharted territories of destruction," with a fivefold increase of climate-related disasters over the past fifty years. Weather-related disasters were now costing an estimated $200 million a day.[33] "There is nothing natural about the new scale of these disasters," said UN Secretary-General Antonio Guterres. Pakistan was then suffering a flood catastrophe that he called "climate carnage."[34] So it is happening. The world is warming steadily and dramatically, and the predicted greater extremes and greater frequency of severe weather are coming to pass.

And we don't just detect it through measurements. The effects are already getting so pronounced that we are now experiencing them in noticeable ways: noticeably warmer summer nights, warmer winter days and nights, intense heat waves, drying soil and vegetation, and intense forest fires and wildfires everywhere from the western United States to Australia to Spain, France, Italy, and coastal Greece. They have caused near unprecedented havoc.

In addition, sea levels are rising. Decades ago, in the 1980s, climate scientists predicted that a warming planet would become most evident first in the Arctic regions. Because several feedback systems accelerate any

Arctic warming, the Arctic would warm much faster than the temperate areas where most people live. That clearly is now happening. It is dangerous because the polar regions help moderate warm temperatures for the rest of the planet, and the Arctic warming can change planetary weather patterns in unpredictable ways.

The warmer temperatures in the Arctic are now having their effect on temperate areas where most people live. The reduced north–south temperature gradient is contributing to stronger storms over specific areas, like the Upper Midwest, the Great Lakes Region, and the Northeast: Big storms over these areas that used to be quickly pushed away and not affect any one area too badly are now able to linger over a region and drop enormous quantities of rain or snow.

Glacier National Park used to have 130 glaciers. When Pollack spoke, fifteen were left. That's the number given for today, although they are smaller. In the not-too-distant future, there may be none. My wife and I visited this beautiful national park in the second week of September 2022, crossing over the high Going-to-the-Sun Road twice. Despite a new snowfall one morning on some higher peaks, actual glaciers were few and far between.

An even more telling effect of a warming climate became literally visible to us. Haze from the multiple forest fires in states to the west obscured the clarity of views within Glacier National Park. Worse, we left and drove south from Whitefish, in northwestern Montana; past the long Flathead Lake; through and past Missoula, Montana; and all through the forests of southwestern Montana and the wide grassy valleys of eastern Idaho to Idaho Falls. Thick smoke and haze from those fires cast a dark pall over the entire region. The area shrouded by smoke extended at least four hundred miles from north to south and I assume at least a similar distance east to west. Any landscape features were obscured. Local people we talked to about it seemed sad and resigned. There was an apocalyptic feel.

I was in the Arctic in late February 2018, at a tiny village north of the Arctic Circle in Alaska called Bettles. There was a lot of snow, and it was cold—but not as cold as it should be. The owner of our lodge was a typical Alaskan, hardy, proud, self-reliant, a bit crusty, and I would guess politically conservative. He had lived year-round in Bettles for eleven years. He told us that he has noticed that northern Alaska is getting much warmer and wetter. No one in our group prompted him to talk about this;

he brought it up himself. While we were there, the temperatures hovered mostly between -17.8°C (0°F) and -23.3°C (-10°F). That is warm for the Arctic in February. "It should be 30 or 40 degrees below zero this time of year," he told us.

When we returned to Anchorage, this front-page headline across the top of the *Anchorage Daily News* greeted us: "Storms Pummel Bering Sea Islands after 'Crazy' Ice Melt-Off." Here are the opening paragraphs:

> A swath of sea ice the size of Minnesota vanished from the northern Bering Sea as warm storms pummeled the region this month, causing ice coverage to fall well below a record low set 17 years ago, scientists say.
>
> About half the ice in the Bering Sea disappeared during a two-week period that ended Saturday, said Rich Thoman, climate scientist with the National Weather Service in Alaska.
>
> The unusual conditions have stunned residents from the region, who say they've never seen so much water around their islands into February.
>
> "(Now) it's just all open water, between this island and Siberia," said Edmond Apassingok, 54 and a hunter from the village of Gambell on St. Lawrence Island. "It's undescribable. It's crazy. Time is broken."[35]

The Antiscience Attacks

So it is against this backdrop of a noticeable warming world that I consider the antiscience movement that has so notably resisted this realization. The situation we have is like looking across a gap of misunderstanding as wide as the Grand Canyon. The gap is wider than the solar system's biggest canyon, Valles Marineris on Mars, four times deeper and five times longer.

On the one side, we have decades' worth of scientific measurements and observations of both increasing atmospheric greenhouse gas levels and a warming climate that is the consequence of that. We have a solid theory, a self-consistent and strongly affirmed scientific explanation, for why that is happening. And we have evidence of our own senses all around us that the world is warming and weather is more frequently going to more extremes.

On the other, we still have people who reject all that—who toss all that aside as easily as they flick away a fly. "It's all a hoax," they claim, repeating what they've probably heard from certain right-wing radio broadcasters every day for decades. Or if they don't go that far, they nevertheless make clear their doubt about the evidence and their distrust of the scientists who have so carefully amassed it. It is an incredibly unhealthy thing for our democracy. But it is part of a general deterioration of any common consensus that we've all seen, and a new and worsening solidification and ever-widening polarization of political and cultural viewpoints. The general decline in civility and in willingness to listen to arguments and evidence we find inconvenient has infected and poisoned our attitudes toward certain areas of science many have now been culturally taught to distrust.

I earlier mentioned graphs of past and recent temperatures from the relevant government science agencies. When the graph is compressed to fit 400,000 or 800,000 years' worth of data onto one page or one screen, the recent upward rise appears as a straight-up vertical line. Reduce the scale to just the last millennium to give more space to recent decades, and it is still a steadily upward-trending line. Climate scientist Michael Mann and a colleague came up with a famous "hockey stick curve" analogy in the 1990s, where a hockey stick lies on its side, pointing upward. It dramatically conveys what seems to be happening. Visual images like this have strong communicative power. Al Gore picked up the analogy and used it in his talks and his famous documentary film (and book) *An Inconvenient Truth.*[36] Such powerful analogies can also attract the scorn of opponents to the idea that the Earth is warming. And this one did. Mann came under harsh attack.

If it could be proved wrong, then the case for global warming could be seriously compromised. If a scientist could show it was wrong, then they could win immediate acclaim. Dozens of scientific groups carried out their own studies to try to confirm or disconfirm the idea. Challenges have been published in leading scientific journals. "Yet a decade and a half later," Mann says, "a veritable hockey league of studies confirms the basic hockey stick conclusion. The most comprehensive study to date has yielded a curve that is virtually indistinguishable from the original 2016 hockey stick curve. The basic finding has stood the test of time."[37]

In late summer of 2022, marked by the horrible extreme weather events all over the globe, I asked Michael Mann (just then moved from Penn State to the University of Pennsylvania) for his most up-to-date assessment of the climate. Here is what he told me: "The appropriate point would be that the IPCC, based on more recent studies that extend the record back further, has now replaced the 'Hockey Stick' with a 'Scythe': The 'handle' is much longer, the 'blade' is much sharper, driving home the truly unprecedented nature of human-caused warming."[38]

The Campaign to Foment Distrust

So the conclusions about a warming climate have become ever more steadily reinforced as each new year's observational data come in. Yet the antiscience attacks started early on and haven't really let up. How did this happen? How did we get to this situation where distrust of commonplace—yet very important—scientific results has become such a serious problem? It did not happen by accident.

The first salvo may have come in 1989. Three scientists known for their conservative views wrote a report for the George C. Marshall Institute questioning the then growing evidence for global warming. The Bush White House and national media picked up this report, and its claims and criticisms were repeated over and over, including by the president and vice president. The resulting public debate became portrayed as a "scientific debate," but it wasn't a scientific debate at all. It was misinformation. At its root, it was an intentional campaign of misinformation. Who would be behind such a campaign, and why?

Well, a handful of hawkish scientists, such as Frederick Seitz and S. Fred Singer, had joined up with think tanks, private companies, and Washington lobbying firms to sow misinformation and distrust about the scientific evidence on hot-button contemporary issues. Global warming was the latest. These scientists had held noteworthy public positions. Although they hadn't been climate researchers, they were both physicists. They had all served at high levels in science administration. Fred Seitz was a former president of the National Academy of Sciences. Fred Singer was the first director of the National Weather Service. Both were associated with the Marshall Institute, which had been established to defend Ronald Reagan's Strategic Defense Initiative ("Star Wars").

Seitz and Singer were using the same tactics they had used earlier to try to undermine the scientific evidence about the health dangers of tobacco smoking and the ozone hole in the atmosphere. From 1979 to 1985, Seitz directed a program for the R. J. Reynolds Tobacco Company that distributed $45 million to scientists to do biomedical research to discredit the findings of the harm of tobacco and defend the company's product. In the mid-1990s, Singer was coauthor of a report condemning the Environmental Protection Agency's (EPA) assertion about secondhand smoke. Earlier he had attacked the US surgeon general's conclusion that secondhand smoke was dangerous. Singer's anti-EPA report was funded by a grant from the Tobacco Institute, but it was channeled through another group, the Alexis de Tocqueville Institution.[39] Just a couple examples of their techniques against climate science follow.

In a letter to *Science* in 1996, Singer criticized the then most recent IPCC report of the world's climate scientists with a series of claims that were incorrect. His claims were not supported by the data, and he had created a straw man, alleging that the report claimed that global warming was the "greatest global challenge facing mankind" while those words did not appear in any of the IPCC documents. As explained in Naomi Oreskes and Erik M. Conway's *Merchants of Doubt*, "Singer was putting words in other people's mouths—and then using those words to discredit them."[40]

Seitz, too, was on the attack in 1996, with a letter to the *Wall Street Journal* in June accusing one of the report's contributors of misconduct. These charges were refuted, but the damage was done. The IPCC's letter to the *Wall Street Journal* giving details of Seitz's mistakes and misstatements was edited down to a few general statements, and this prompted the boards of both the American Meteorological Society and the University Corporation of Atmospheric Research, leading organizations for atmospheric research, to protest and themselves publish the omitted parts of the response. Their letter referred to a "concerted and systematic effort by some individuals to undermine and discredit the scientific process."[41]

Forgive me two personal notes here, as an aside. My first job in Washington, DC, fresh out of graduate school at Columbia, was a three-year stint as editor of *News Report*, then the monthly newsletter of the National Academy of Sciences. The academy's president during those three years was Frederick Seitz. I knew he had been a solid-state physicist (researching semiconductor materials) at the University of Illinois. He was thin, bald (even then), scholarly looking. He seemed mild mannered. I never heard any criticism of him. In fact, official notice of my first-ever raise came in a personal letter to me signed by Fred Seitz; I was cleaning out old files in my garage in 2020 when I came across the original letter.

When I was at the National Academy, I attended regular meetings of the Academy's Council, where Seitz was present, but I never got any sense that he was especially politically oriented or had some kind of agenda against certain scientific findings. That, I am sure, would have been anathema to the academy. I forgot about him until years later when sitting at my desk at Sandia National Laboratories in the 1990s, I began seeing a series of op-eds by a Frederick S. Seitz in the *Wall Street Journal* strongly (and in my view a bit strangely) critical of climate science. It took me awhile to realize that this had to be the same Seitz I had known so many years earlier. I couldn't believe it. It didn't seem plausible. Seitz had long ago retired from the National Academy of Sciences, but he was now using the prestige of that past position to lend seeming credibility to his increasingly antiscience views.

Likewise, in all my years as a working science journalist attending scientific meetings around the country (such as the big AAAS annual meetings), Fred Singer was often there. He always seemed a bit apart from the rest of the scientists, and when he spoke up at question sessions, he seemed always to be voicing a view contrary to what had just been said. I didn't quite get it, but it was clear he viewed things differently from most of his scientific colleagues. He seemed not to be accepted by them. I always thought he looked lonely and a bit sad.

Little did I know then that Seitz and Singer were involved—or if not then, they soon would be—in the now notorious campaign to support big tobacco against the biomedical science that demonstrated the detrimental health effects of smoking—and they were also closely aligned with the fossil fuel industry and its think tanks to undermine the science about climate change.

These think tanks and corporations were soon joined by a number of foundations, lobbyists, and then the fossil fuel industry itself to mount a coordinated campaign to discredit the very idea of global warming and the findings of climate scientists supporting it. They wrote op-eds and letters to the editor, they distributed negative papers, they made public accusations that were widely distributed, they held their own conferences, and they weren't that interested in presenting any scientifically accurate or balanced view. They had an agenda.

These two men and a handful of others equally well situated, like Robert Jastrow, the astrophysicist and popular science book author, were working steadily to subvert certain scientific findings, and they were good at it. They were well known; they were savvy in the workings of Washington; they attracted media attention; they weren't shy about pointing out their scientific credentials; and with powerful corporations and politically oriented think tanks and well-funded foundations supporting them, they were effective. Very effective indeed. They and their cronies carried out countless "dirty tricks" against climate scientists.[42]

Scientists with the major oil companies were well aware of the evidence for human-induced climate change from the burning of oil and other fossil fuels. Researchers and engineers for Exxon and Imperial Oil were quietly incorporating these climate projections into their own plans, knowing they had to adapt their operations. Yet the companies' management kept all these internal concerns quiet. For all external audiences, they maintained that the science was too murky to warrant action. This was all documented in great detail in October 2015 in a major investigation by the Energy and Environmental Reporting Project of the Columbia University Graduate School of Journalism and the *Los Angeles Times*, who published it.[43]

Historians of science Naomi Oreskes (now at Harvard University) and Erik M. Conway fully document the long campaign to discredit climate science in their marvelous book *Merchants of Doubt*. It describes in delicious detail the political campaigns Seitz, Singer, and their colleagues and supporting organizations carried out against the science about tobacco, acid rain, the ozone hole, and then global warming. "On every issue [these men] were on the wrong side of the scientific consensus," Oreskes and Conway write.[44]

The whole campaign against the science of global warming then and now was the same: fighting facts, merchandising doubt, and misrepresenting great scientists—as Singer later did in attaching Revelle's name (when Revelle was eighty-two years old and near death) as coauthor without permission to one of his own papers, making it appear that Revelle agreed with his views. That's as unethical and disreputable as it gets. He then managed to quash an effort by another scientist to clear Revelle's name. More confusion was sowed. Antiscience warriors used the same misinformation tactics that had been so effective for decades on behalf of the tobacco industry: sow doubt ("doubt is our product").

When people die, they usually get laudatory obituaries. Singer's role in the campaign against climate science was so notorious and scurrilous that when he died on April 6, 2020, at the age of ninety-five, one science/environment reporter, Paul Thacker, cast all convention aside. In a piece subtitled "Why Speak Well of the Late Climate Denier Fred Singer, Who Spent over a Half Century Attacking Credible Science and Scientists," Thacker outlines at length Singer's subterfuges. He then recalls finding himself in the same restaurant with Singer one day and couldn't even eat his meal: "They say you shouldn't speak ill of the dead, but I spoke ill of Singer that day, and I feel no need to stop just because the bastard doesn't breathe. What I saw that day was the face of evil."[45]

Thacker adds that he wasn't alone in that view: "Many have said to me in private that they also found him evil." He quotes writer Ann Lamott, "If people wanted you to write warmly about them, they should have behaved better."[46]

It's Human Nature: Does Evidence Matter?

When organized forces intentionally cast doubt on a whole field of science using deception and distortions, that is obviously a bad situation. It is antiscience at its worst. But there's more to it than that. There is more than these organized antiscience efforts to explain why climate science became such a hot-button issue. It is not controversial among most scientists at all, but it remains controversial among some of the public. And that goes to show how our minds work when faced with conflicting information and how we tend to group ourselves with people who share our important cultural views and outlooks.

Studies carried out by a new wave of social science and communications science researchers show that facts and evidence hold very little sway over us when it comes to matters that seemingly challenge our deep-seated beliefs or values. Here we take a variety of mental short cuts. We tend to ignore evidence that conflicts with our beliefs, and we tend to seek out evidence that reinforces our beliefs (confirmation bias). We listen to news channels and commentators and analysts we trust and ignore those we don't. We now seem to accept the science we like, and doubt the science we don't like. We do the same in our personal interactions every day, and all this eventually divides us into increasingly homogeneous groups. And it keeps us from seeing the other side of important issues and considering whether we might be wrong.

We also are tribal, a trait going back to the dawn of human history. When it comes to what we consider most important, we want to be aligned with those who are most like us, who believe the same things we do, and who act the same ways we do (maybe even who look the way we do). This might be an admirable trait for fighting against a common foe. It is less so when it is really in our best interests to welcome new thinking—when faced with new information and new insights that we need to understand and absorb to avoid future disasters or even catastrophes. People listen to whom they trust—family, friends, and public figures they like and admire. They don't listen to people they don't trust, no matter how knowledgeable or how well informed those people might be.

So it is ever more important for all those who support good science to not only speak out about the facts of their science but also to do so in ways that seem respectful and appealing to those who might otherwise distrust them and tune them out. This is difficult. It's not part of most scientists' training. But in controversial areas that challenge core beliefs, it is necessary. Scientists and scientific groups concerned about this problem now must do more than just foundational, groundbreaking research—as if that's not enough! They now also need to present the results and the possible consequences of it in ways that seem comfortable and inviting, not discomforting or threatening. It's not fair, but it seems to be the new reality.

CHAPTER TEN

THE RISE OF ORGANIZED SKEPTICISM

To begin, I must take you back some decades to the mid-1970s. Three extraordinary years from 1974 to 1976 marked the beginning of an organized skeptical movement. And by that, I mean the first systematic and sustained permanent effort to publicly examine and critique a wide variety of extraordinary popular claims ignored by or unsupported by science.

The world back then was awash in unexamined paranormalism. Astrology was in high vogue (the "Age of Aquarius"). "What's your sign?" passed for a mainstream conversation starter. Psychics and belief in psychic powers reigned everywhere. Some well-publicized parapsychological experiments had convinced certain scientists of the reality of psychic powers (psi). Uri Geller was bending cutlery and fooling even some prominent physicists into thinking he was doing it by the sheer power of his mind. Flying saucers and UFOs were penetrating our skies and enticing strong belief in extraterrestrial visitations. Ships and airplanes were seemingly disappearing daily into the Bermuda Triangle. Birth-date-based biorhythms were a big thing. Crystals and pyramid power had ardent New Age followers. Erich von Däniken was writing a series of phenomenally best-selling books about *Chariots of the Gods* and ancient astronauts and the near-racist notion that ancient edifices of the Americas had to have been built by aliens and not by the skill and hard work of the people who lived there at the time.

But a scientific response to all this credulousness was beginning to brew. In February 1974, the American Association for the Advancement of Science (AAAS), at its annual meeting in San Francisco, conducted a most extraordinary symposium called to examine the claims of Immanuel Velikovsky and give him the scientific hearing he had long demanded. Velikovsky's *Worlds in Collision* book had gone through seventy-two printings since 1950 and garnered both the wrath of scientists and the science-snubbing approval of some prominent literati.[1]

Five hundred scientists filled the auditorium that day. I sat in the front row while the seventy-seven-year-old Velikovsky, his silver hair and demeanor lending him an almost biblical air, and the passionate young astronomer Carl Sagan, already world famous at age thirty-nine, squared off against each other: two amazingly charismatic figures in collision. It was electrifying. I was one of the science writers who quickly ran over to Sagan afterward and persuaded him to loan his fifty-six-page manuscript to us temporarily so we could take it to the AAAS press room for photocopying, allowing us to all have copies. As I recall, it was double-spaced and filled with scores of changes and emendations Carl had made, right up to delivery, in his own hand.

Sagan's ten-part dissection that day of Velikovsky's arguments is debated still today, and the subject stimulates strong controversy and intense philosophical interest even now. I deeply admired Sagan's willingness and determination to take on the ideas of one of the era's most influential and intelligent pseudoscientists, to read Velikovsky's books carefully and to write such an extended critique and present it in such a public forum. Until then, few scientists were willing to do such things. To my mind, this was a seminal moment in establishing a new tradition of scientists and scientific-minded scholars publicly committing time and effort to examining—not ignoring—extraordinary claims of a pseudoscientific, fringe-scientific, or paranormal nature.

Later in 1974 came noted aerospace journalist Philip J. Klass's book *UFOs Explained*.[2] It patiently, thoroughly, and in considerable detail examines and clearly explains many of the most prominent UFO and flying saucer claims. The next year, things really got into high gear. Larry Kusche published *The Bermuda Triangle Mystery Solved*, still regarded as one of the classic debunking books of all time.[3] (Debunking is the *end result*. Kusche dislikes being characterized as a debunker; he just wanted to find out the truth, and he did.) That same year, magician/investigator James "The Amazing" Randi published his book *The Magic of Uri Geller*, the first exposé of Geller's methods and tricks, in itself a landmark achievement.[4]

And then in the September/October 1975 issue of *Humanist*, at the instigation of its editor, the philosopher Paul Kurtz, came "Objections to Astrology." This short proclamation was a clear statement straightforwardly cautioning the public "against the unquestioning acceptance of the predictions and advice given privately and publicly by astrologers."[5] Respected University of Arizona astronomer Bart J. Bok wrote the two-page statement with the help of science writer Lawrence Jerome and Kurtz, a professor of philosophy at the State University of New York at Buffalo. They got 186 leading scientists, including 19 Nobel Prize winners, to sign it, giving the statement instant public credibility. Accompanying it were two supporting articles, including Bok's "A Critical Look at Astrology."[6] Such was the public appeal of astrology at that time that this fairly mild statement of objections ignited a firestorm of publicity and controversy. Prometheus Books quickly republished the whole package as slim booklet.[7]

Something big was happening. The scientific examination of widely accepted claims that to scientists were at best misguided and at worst bogus was now a major news story and a part of the new cultural debate.

First Steps

In the meantime, magician/investigator James Randi, psychologist Ray Hyman, and noted author/critic Martin Gardner had been discussing among themselves the need for some kind of organization that could address such claims. Gardner was a hero already to many for his classic book about pseudoscientific and crank claims, *Fads and Fallacies in the Name of Science*, and for his popular and clever Mathematical Games column in *Scientific American*.[8] They corresponded among themselves about the possibilities.[9] They considered and rejected various names for their envisioned organization. First was Sanity in Research (SIR), a kind of play on the initialisms of the Stanford Research Institute (SRI, later Stanford Research International), the then recent source of several notorious claims about remote viewing and other psychic powers. Then, in 1974, came the Committee for Constructive Skepticism and, by September 1975, Resources for the Scientific Evaluation of the Paranormal (RSEP). None of these names stuck. But there was a bigger problem than what to name the group: Hyman, Gardner, and Randi had no experience or skills in creating or running organizations.

"I am still struggling with various alternatives and issues of justification," Hyman wrote to Gardner in December 1974. "I hate to get involved in something so huge without some prior experience."[10] By August 1975, all seemed lost. "I've come to the conclusion . . . the proposed organization is a fruitless endeavor," Gardner wrote to Hyman. "There seems to be no way to get it funded, no way to exclude the nuts, and nobody has the time and money to invest in handling correspondence, etc. . . . As I see it now, the best plan is for those of us were [*sic*] skeptics to keep making our voices heard now and then, and forget about any formal organization."[11] They needed help.

Enter Paul Kurtz. Kurtz's experience with "Objections to Astrology" had impressed on him the need for scientific and scholarly examinations of paranormal and fringe-science claims that have wide popular appeal. Kurtz was concerned about this uncritical acceptance and the lack of scholarly rebuttal. He saw that there was great public interest in such critiques. Kurtz was even initially open to the possibility that some claims of telepathy, neoastrology (a set of specific claims that differed from classic astrology), and ufological visitations might have at least some possible validity, but he was eventually persuaded otherwise as he learned better evidence.[12]

But most important, Kurtz was an organizational genius. He had already founded and was president of Prometheus Books and later explained, "One reason I founded Prometheus Books in 1969 was that I thought the country needed a dissenting press which would defend a thoroughgoing naturalistic and scientific rationalist outlook."[13] Over the next four decades Prometheus would publish hundreds of skeptical and scholarly books examining paranormal and fringe-science topics, promoting good science, and criticizing the tenets of religion.

By now Kurtz was in touch with Marcello Truzzi, a professor of sociology at Eastern Michigan University. Truzzi had become deeply interested in the sociology of paranormal claims and beliefs. He knew some of the paranormalists and some skeptics. Truzzi had corresponded a bit with Randi, Hyman, and Gardner. Sometime in late 1975 or early 1976, they all got connected.

"The New Irrationalisms" Conference

In early 1976, Kurtz, Truzzi, and Lee Nisbet, one of Kurtz's recent philosophy PhD students, began to organize and eventually publicize a unique conference at the brand-new suburban Amherst campus of the State University of New York at Buffalo. (It would become the first conference ever held on that campus, Kurtz later said.) They titled it "The New Irrationalisms: Antiscience and Pseudoscience."

I recently spoke on the telephone to Nisbet about those events four and a half decades ago. He recalls it as a "very exciting time." He got thrown into the task with no preparation, but Kurtz had a way of just going ahead and getting things done. "You didn't have to have experience—just do it was the idea. It was liberating," Nisbet says, "because it relieved the pressure of making mistakes." Letters went out to would-be speakers. Requests were sent to a group of outstanding scientists and scholars seeking their participation in the organization. Kurtz and Nisbet solicited some scholars with a form letter, but Kurtz personalized many of the letters, Nisbet says. He still has some copies.

"I drafted a call inviting a number of leading scholars to the inaugural session," Kurtz would recall a quarter-century later. "This was endorsed by many leading philosophers, including W. V. Quine, Sidney Hook, Brand Blanshard, and Antony Flew. And I invited many of the leading skeptical critics to the opening session—Martin Gardner, Ray Hyman, Philip J. Klass, James Randi, L. Sprague de Camp, and Milbourne Christopher."[14] And in that recollection, he acknowledged that one of the reasons he decided to create a new organization was his experience with the "Objections to Astrology" statement, "which received widespread attention. And I surmised that surely what we did about astrology could apply to all these other unchallenged fields."[15]

Sitting at my editor's desk on the third floor of *Science News*'s redbrick Civil War–era building overlooking historic N Street in Washington, DC, a half-mile north of the White House, in mid-April 1976, I vividly remember reading the mailed invitation to cover that conference. It promised international attention and controversy. They said they would announce the formation of a committee to "examine openly, completely, objectively, and carefully . . . questionable claims concerning the paranormal and related phenomena, and to publish results of such research."

The invitation intrigued me. It spoke of an "enormous increase in public interest in psychic phenomena, the occult, and pseudoscience," much of it accepted uncritically and without scientific rebuttal. It pointed to the need for some "strategy of refutation." It added, "Perhaps we ought not to assume that the scientific enlightenment will continue indefinitely, for all we know . . . it may become overwhelmed by irrationalism, subjectivism, and obscurantism." It called on the scientific and educational community to respond "in a responsible manner—to its alarming growth."

I had recently become deeply interested in such issues, and we at *Science News* had struggled with them in three articles written in response to reader requests. We'd had only partial success. Martin Gardner, in fact, complained about their inadequacies in 1974 in a personal letter to me, as editor of *Science News*. I answered him that we in the media needed the help of informed experts like him to get reliable scientific information about these kinds of claims because they don't go through the normal kinds of scientific scrutiny. We and all journalists needed help from informed scientific experts. Also, I remembered the "Objections to Astrology" statement, which we had reported on in *Science News*, and the surprising controversy it had generated worldwide. I knew I had to go. I flew up to Buffalo.

Held April 30–May 1, 1976, the conference brought together the leading lights of scientific skepticism to air their scientific viewpoints about all the bizarre claims then in the wind. Like the AAAS Velikovsky symposium, it was electric. I had covered all kinds of scientific conferences before, but this one was way different. None had dealt with the things *these* scientists, scholars, and investigators were examining: people's deepest interests and emotional passions and intellectual misperceptions on a wide variety of topics that fascinate the public, including but not limited to psychics, UFOs, astrology, bogus archaeology, the Bermuda Triangle, Kirlian photography, and claims of life after death.

Kurtz spoke passionately at the conference on the scientific attitude in contrast with that of pseudoscience and antiscience. He referred to "cults of unreason and other forms of nonsense inundating even supposedly advanced societies." He gave examples of the "current rejection of reason and objectivity" and worried that "large numbers of people are apparently ready and able to believe in a wide variety of things, however outrageous, with sufficient proof or evidence." He called on scientists to be willing to investigate claims of new phenomena that interest the public. He lamented a "tendency for the credulous to latch onto the most meager data and frame vast conjectures, or to insist that their speculations have been conclusively confirmed, when they have not been."[16]

"Often the least shred of evidence for these claims is blown out of proportion and presented as 'scientific' proof," Kurtz said in his opening address. "Many individuals now believe that there is considerable need to organize some strategy for refutation. Perhaps we ought not assume that the scientific enlightenment will continue indefinitely; for all we know, like the Hellenic civilization, it may be overwhelmed by irrationalism, subjectivism, and obscurantism."[17]

Truzzi also spoke, cautioning scientists not to place all occultist groups into one package. He offered a taxonomy of occultism, placing claims along a five-point scale according to whether their sources of validation were scientific, mystical, or something in between. He also emphasized that what distinguished science from pseudoscience is not subject matter but methodology, listing such principles as falsifiability and replicability. Truzzi proposed two additional principles that have been the hallmark of skepticism ever since: "First, the burden of proof is on those who claim the existence of an anomaly; second, extraordinary proof is necessary for extraordinary claims."[18]

Committee for the Scientific Investigation of Claims of the Paranormal Founded

And it was here that the Committee for the Scientific Investigation of Claims of the Paranormal (CSICOP; now shortened to the Committee for Skeptical Inquiry) was founded. At a well-attended news conference, Kurtz offered,

> We wish to make it clear that the purpose of the committee is not to reject on a priori grounds, antecedent to inquiry, any or all such claims, but rather to examine them openly, completely, objectively, and carefully.
>
> We do not yet know how large our committee will become or how ambitious its efforts will be. . . . We have invited leading scientists and experts in many fields to join us in this important endeavor.[19]

National news media covered the event, including the *New York Times*, *Science* magazine, *Time*, and my own cover article in *Science News*.[20] The article generated more reader interest, measured by number of letters to editor received, than any other *Science News* article in our memory. That astonished us. Something was in the air. "Surprising to me was the fact that CSICOP and the *Skeptical Inquirer* took off from day one and became an instant success," Kurtz recalled later. "The media found us so fascinating that they began to call us constantly for information and have not stopped since."[21]

Nisbet remembers it the same way. "It took off beyond our wildest dreams," he told me in 2021. Nisbet became executive director of the fledging organization and soon was overwhelmed with media inquiries, correspondence, and other duties, but he got it done. "Paul would listen," he recently told me. "We had strategy planning sessions. He respected you if you could do the work. He was attracted to people who were intellectually assertive."

Nevertheless, not all the reaction was positive. "Although we received a warm reception by mainline science magazines," Kurtz recalled, "we were bitterly attacked by believers. They accused us of being 'the gatekeepers of science.'"[22]

Kurtz, Randi, Gardner, Hyman, and Truzzi were on the original CSICOP Executive Council. So were Phil Klass and Nisbet. Carl Sagan was one of the founding fellows. So was Isaac Asimov. So was famed behavioral psychologist B. F. Skinner. So were a number of other famous scientists and scholars (including, by my count, ten professors of philosophy, among them Nagel, Quine, and Flew).

From the beginning CSICOP had three strategies, described by Kurtz: (1) Scientists should devote some research effort to investigating the paranormal claims that interested the public; (2) researchers should bring the results of their studies to the public ("We were eager to encourage further research and to publish the findings."); and (3) "our long-range goal was public education of the aims of science, particularly an appreciation for scientific methods of inquiry and critical thinking."[23]

From *Zetetic* to SI

Truzzi became the first editor of CSICOP's journal, initially published just twice a year and called the *Zetetic*. There was soon a fallout between Truzzi and the CSICOP Executive Council, and Truzzi departed the next year. The disagreements revolved around two main issues: Truzzi wanted the magazine to be a very scholarly, academic, sociology-style journal; the rest wanted it to appeal more to the public and have a much wider and broader audience. Truzzi also wanted to bring paranormal proponents into the organization and give them equal time; the others strongly disagreed. They felt paranormalists already had the public's ear and plenty of outlets for their claims, whereas the scientific viewpoint was seldom heard and should be the organization's focus.

In August 1977, at a meeting of CSICOP at the old Biltmore Hotel in New York City (I'd been invited to speak to them on a science editor's view of reporting on the paranormal), the Executive Council, in executive session, appointed me to succeed Truzzi as editor. At the same time, they elected me a fellow and member of the Executive Council.

As for me, I at first greatly admired Truzzi. I liked his knowledge and forthrightness. I felt horrible that my invitation to become editor of the magazine came at his expense. But I soon came to see the problem my colleagues had noticed. It seemed he actually felt that the best way to resolve disputes about the validity of various paranormal claims was to just talk about them—to give equal time to proponents and critics to argue them out. But scientific claims aren't adjudicated by who is best at arguing their position or how much verbiage they can produce. It is not felicity with words that matters; it is a hard-nosed assessment of the scientific evidence for or against a claim. Truzzi himself, ever the sociologist, loved to argue at length, and he did so in person, in his own publications, and in voluminous correspondence. His arguments back and forth about such matters with Martin Gardner in hundreds of letters over the years would eventually fill an entire book, *Dear Martin, Dear Marcello*.[24] These and other disputes with Truzzi continued to occupy CSICOP's time and energy for the next decade because once he departed the organization, he became a public critic. To us, it seemed he was far more concerned about fairness, especially to paranormalists, than what the scientific evidence showed.

I have edited the magazine ever since that day in August 1977. The very next year, in 1978, we renamed the magazine the *Skeptical Inquirer* and made it a quarterly. Then in 1996, at my instigation and after polling readers for what they wanted, we increased it to bimonthly publication and reformatted the magazine from its original digest size to a standard magazine.

The *Skeptical Inquirer* has always addressed myriad subjects, but Kurtz felt that three major ones were the most important in CSICOP's earliest decades: astrology (including neoastrology), parapsychology (including Uri Geller), and ufology. These three broad subjects occupied everyone's considerable time and effort. They all created controversy, with lots of counterclaims fired back from unhappy proponents of the claims. But CSICOP held firm, and at least some people were persuaded to a more scientific viewpoint about them. The committee and the magazine also dealt with other "weird" claims (Kurtz's adjective), and in 2001, he expanded on four: communicating with the dead, miracles, intelligent design, and alternative medicine.[25]

There is, of course, a long record of skeptical activity preceding CSICOP's founding in the 1970s. One earlier organization should be mentioned: Belgium's Comite Para. It was formed in 1949 and when CSICOP was founded had been in continuous existence, but I don't think it influenced CSICOP's creation. I could also mention Harry Houdini's many exposés of spiritualists and would-be psychics (the title of his book *Miracle Mongers and Their Methods: A Complete Exposé* says it well) and quackbuster physician Arthur C. Cramp's spirited and systematic examinations and exposés of medical quackery for the American Medical Association throughout the first third of the twentieth century.[26] Other individual efforts go far back into history. If you think *I'm* taking a long perspective, then I recommend reading my skeptic colleague Daniel Loxton's detailed article.[27] A new book that also takes a very long historical perspective on exposing fraudulent claims is my colleague and former psychiatry professor and skeptic Loren Pankratz's *Mysteries and Secrets Revealed: From Oracles at Delphi to Spiritualism in America*.[28]

But for the English-speaking world, CSICOP was the first well-organized and broadly constituted organization established on a permanent basis to examine paranormal and fringe-science claims from a scientific viewpoint—and to serve as a forum for scholarly discussion of all the relevant academic, social, and educational issues surrounding them. The founding of CSICOP can be considered the start of the modern skeptical movement.

Twelve Big Topics

For my part, I believe that in its first several decades CSICOP and the *Skeptical Inquirer* made a difference in at least twelve major popular areas that had widespread popular appeal. By making a difference, I mean we were effective in educating the media and the public about the topics' scientific shortfalls and the need for being more circumspect when reporting on them. The first ten topics are cold reading (a technique psychics use to deceive people into thinking they know secret things about them); "serious" astrology (not the popular version of astrology columns, although neither has scientific validity); fire walking; full-Moon claims; near-death experiences; subliminal persuasion; false memories; Satanic panic claims; facilitated communication; and reincarnation. These ten are all the subject of notable investigations and articles in the *Skeptical Inquirer* that got considerable public attention. We can add to those topics three more (making a baker's dozen): the hundredth-monkey phenomenon (see later in this chapter); James Randi's famous Project Alpha experiment (placing two young sham "psychics," actually magicians, in the McDonnell Laboratory for Psi Research at Washington University to see if the researchers could detect their deceptions; they didn't); and the testing of psi claims in China (particularly claims of qigong energy and supposed psychokinetic powers of a group of Chinese children in Xian; CSICOP sent Randi and a delegation of the Executive Council, myself included, to China to investigate in person). All were notable investigations. I think it is fair to say that they all had some positive effect in dampening credulous enthusiasms for sensational, popular claims.[29]

Just one example of what I mean: In the mid-1980s, philosopher Ron Amundson investigated for the *Skeptical Inquirer* a widely publicized New Age claim by writer Lyall Watson concerning apparently new behaviors of Japanese macaques (a type of gregarious monkey) on two Japanese islands. The claim was that, even on different islands, the macaques all learned a particular potato-washing behavior once a critical mass of the population (one hundred) learned it. This was supposedly a paranormal "group consciousness"—an extraordinary claim that, if true, would seemingly have widespread implications.

Amundson checked the scientific sources Watson cited. Just these papers alone themselves invalidated his claims. There was no hundredth monkey. Amundson wrote about the flaws in Watson's article in some detail. He pointed out that the way belief in this claim grew exemplifies the processes of pseudoscience. Amazingly, Watson later admitted he had made up most of the details: "I accept Amundson's analysis of the origin and evolution of the Hundredth Monkey without reservation. It is a metaphor of my own making, based—as he rightly suggests—on very slim evidence and a great deal of hearsay."[30]

When we included these two reports by Amundson in a 1991 anthology of forty-three selected *Skeptical Inquirer* articles and essays (including by Carl Sagan and Isaac Asimov), we decided, with Amundson's permission, to title the whole volume *The Hundredth Monkey*.[31] It was very successful and is still in print.

Skeptical Inquirer Anthologies

The tradition of collecting prominent *Skeptical Inquirer* articles in occasional book-length anthologies started only a few years after the beginning. At a lunch with the Executive Council in 1980, I briefly mentioned the possibility to Paul Kurtz. I thought perhaps his Prometheus Books might be interested in publishing the first anthology. "Let's do it," he said. It was typical of his decisiveness. So I quickly began to assemble candidate articles, organizing them, getting the writers' permissions, and writing an introduction. The result was *Paranormal Borderlands of Science*.[32] At 470 pages, it includes forty-seven articles organized into sections on psi phenomena and beliefs, tricks of the psychic trade, Geller-type phenomena, stories of life and death, rhythms of life, exploring science's fringes, astrology, land and sea, extraterrestrial visitors, cult archaeology and biology, planetary pinballs (Velikovskyism), and UFOs. They include original research, critical essays, analytical articles, and investigative reports. They vary in style and approach. "What they have in common," I write in the introduction, "is the application of intelligence, reasoning, and critical inquiry to subjects too often spared from any sort of skeptical, informed attention."[33] The book got some good reviews, including one in *Nature*, and the tradition of publishing books of *Skeptical Inquirer* articles was underway.

Next came *Science Confronts the Paranormal*, the aforementioned *The Hundredth Monkey*, *Encounters with the Paranormal: Science, Knowledge, and Belief*, and *Science under Siege*.[34] We also did a timely single-subject anthology, *The UFO Invasion*, that gathers all of *Skeptical Inquirer*'s most important articles up to that time on UFOs, the Roswell incident, alien-abduction claims, and claims of government coverups.[35] Many of the magazine's investigations on this set of topics came in the 1990s, when there was a flurry of new public interest and attention in these subjects. All these titles, except for *Science under Siege*, are still in print with Prometheus (now an imprint of Rowman and Littlefield). I think most, if not all, these articles stand the test of time and remain valuable references today. As I write, the Prometheus imprint has plans to publish a new *Skeptical Inquirer* anthology, *Unreason*, a collection of forty-five articles from the previous four years.

CHAPTER ELEVEN
SKEPTICISM GOES GLOBAL

Today and Beyond

The *Skeptical Inquirer* continues to publish to this day, with both its bimonthly print magazine (also available digitally, including a digital search function of all back issues) and a website, https://skepticalinquirer.org, that carries many additional original columns and features. Thinking back, I believe I took a middle road between the two original visions of the magazine described in chapter 10: *Skeptical Inquirer* has been scholarly and professional and respectable (in my humble opinion), but we've also attempted to make it readable and appealing and valuable to more general readers. I have often referred to it as a "semischolarly and semipopular" magazine. It bridges the divide between scientists, scholars, and academics on the one side and more general but equally intelligent, caring, and concerned readers on the other. This is an in-between place I have been comfortable with my entire professional career as a science writer and editor.

Paul Kurtz wrote about the beginning of the modern-day skeptical movement often. For a fuller story of this, see especially his 2001 book *Skeptical Odysseys*, with personal essays by thirty-nine other contributors (including me) whom he invited to commemorate the Committee for the Scientific Investigation of Claims of the Paranormal's (CSICOP) twenty-fifth anniversary.[1] He wrote again, in the *Skeptical Inquirer*, on the thirtieth anniversary of CSICOP.[2]

I wrote my own invited 8,400-word history of the first two decades of CSICOP for *The Encyclopedia of the Paranormal*, which is now on *Skeptical Inquirer*'s website.[3] It provides details of the controversies CSICOP went through in its early years, summarizes some of the major coverage by journalists and scientific scholars that CSICOP stimulated, and presents many examples of noteworthy articles published. The full history since then has not yet been written (perhaps this chapter is a start), but for the twenty-fifth-anniversary observance of CSICOP and the *Skeptical Inquirer*, I wrote "CSICOP Timeline: A Capsule History in 85 Easy Steps."[4] And in 2016, in observance of the fortieth anniversary, in the second of two special issues on "Issues in Science and Skepticism" and "Odysseys in Science and Skepticism," we updated it with "Committee for Skeptical Inquiry Timeline, 2001–2016," which lists key events over that ensuing fifteen-year period.[5]

Changing to the Committee for Skeptical Inquiry

Two important changes came in early 2007: CSICOP's Executive Council formally shortened the organization's name from Committee for the Scientific Investigation of Claims of the Paranormal to the Committee for Skeptical Inquiry (CSI). It also revised CSI's mission statement to replace "paranormal and fringe-science claims" with the broader "controversial or extraordinary claims." So the mission statement (short form) now reads, "to promote scientific inquiry, critical investigation, and the use of reason in examining controversial and extraordinary claims." The changes, approved at an Executive Council meeting on September 27, 2006, in Oak Brook, Illinois, were announced and elaborated on in the January/February 2007 *Skeptical Inquirer*.[6]

Our reasons for the changes include our desire for a shorter name, our dislike of still being associated with the word *paranormal*, and our determination to emphasize "critical inquiry, scientific thinking, and the scientific outlook."[7] Our old twenty-syllable name just didn't work anymore in this age of TV and radio sound bites. The joke was that a TV interview would be over before the whole name could be said. Also, many in the group's inner circle had become increasingly uncomfortable with having the word *paranormal* in the name, even if it was clear it meant they were critically *examining* the "paranormal."

Founder and chairman Paul Kurtz wrote an editorial, "New Directions for Skeptical Inquiry," explaining more about the changes. He notes that CSICOP had reached a "historic juncture: the recognition that there is a critical need to change our direction. Under the new name, the Committee for Skeptical Inquiry (CSI) will not confine itself primarily to the scientific investigation of claims of the paranormal—we never have—but will deal with a wider range of questionable claims that emerged in the contemporary world." He called attention to the magazine's subtitle, *The Magazine for Science and Reason*, as the "best description of our overall mission: to explicate and defend the importance of scientific inquiry, thus contributing to the public understanding of science."[8] As Kurtz notes with his inclusion of "we never have," *Skeptical Inquirer* has never really confined itself to only paranormal-related topics. Over the decades as editor, I wrote many columns referring to a broadening of our interests, including issues of science and reason that affect society more generally.

In my editorial on the changes, I note that the CSICOP name was beloved by many of us and would continue to be used for some time in our own minds, but a shorter name was necessary. I also point out that the Executive Council kept *Committee* as part of the name because CSICOP had long also been referred to as "the Committee." I reiterate the broadened mission and conclude, "Implicit in all this is that, as before, we provide quality, informed scientific analysis and deal preferentially, if not always exclusively, with testable, fact-based claims, not assertions based mostly on opinion or ideology."[9]

In January 2015, the Committee for Skeptical Inquiry and its sister organization, the Council for Secular Humanism (CSH; publisher of *Free Inquiry* magazine), formally merged into the Center for Inquiry (CFI), both becoming programs of CFI. There was no change to CSI's mission or to *Skeptical Inquirer*'s content. CFI had also been founded by Kurtz in the 1990s and since then had provided logistical support to CSICOP and to the CSH. In the years since then, the broader CFI organization (which now includes the Richard Dawkins Foundation for Reason and Science and a number of other smaller entities) has provided added stability and supportive organizational infrastructure for CSI and the growth of the magazine, whose total circulation (print and digital) is now about 25,000.

The idea quickly spread. Skepticism was in the air. To return to the 1980s, CSICOP created an information kit and helped found or inspire local and regional skeptic groups in the United States around the same ideas and principles. The Bay Area Skeptics was the first such group, but soon there were dozens of others. Inspired by CSICOP, they were independent and autonomous and excellently situated to deal with local and regional issues. By 1990, the *Skeptical Inquirer* listed independent skeptic groups with aims similar to CSICOP's in Alabama, Arizona (two), California (five), Colorado, the District of Columbia and surrounding areas (National Capital Area Skeptics), Florida, Georgia, Illinois, Indiana, Kentucky, Louisiana, Massachusetts, Michigan (two), Minnesota (two), New York (three), North Carolina, Ohio, Pennsylvania (two), South Carolina, Tennessee, Texas (four), Washington, and Wisconsin.

Going International

Paul Kurtz was an internationalist in both vision and action. He actively encouraged forming similar groups in countries worldwide, and he helped make that happen. Kurtz had a lively mind, tremendous energy, strong intellectual credentials, a high academic position, a global vision, and the bearing and instincts of a diplomat. He traveled frequently to countries all over the world to meet with leading scholars and intellectuals. All these attributes he combined—somewhat incongruously for an academic—with the skills of a first-class promoter, and this made him an effective and very inspiring figure who easily attracted scientists, scholars, investigators, and supporters from all parts of the world to help fight the proliferation of pseudoscience and unreason.

CSICOP and Randi inspired the creation of the Australian Skeptics, which quickly became one of the most active skeptic groups in the world. In 1980, concerned about some prominent paranormal claims there, Australian entrepreneur, aviator, and publisher Dick Smith heard about Randi through the *Skeptical Inquirer* and invited him to Australia to test dowsers—water diviners. Water diviners are interesting because they genuinely and sincerely believe in their abilities to find underground water using some kind of divining stick or rod. Using a system of buried pipes, Randi set up controlled tests. The dowsers failed, and as Randi predicted, they couldn't understand how they had failed because their divining had always worked before. They had never been tested in any real way before this, so those tests got tremendous publicity in Australia and resulted in formation of the Australian Skeptics to carry on such work. That organization has carried out a number of other prominent investigations—including several more with Randi's participation. They've held an annual national conference every year since then (virtual in the 2020 pandemic year) and publish their own lively quarterly magazine, *The Skeptic*. They proudly celebrated their fortieth anniversary in 2020.

The United Kingdom's magazine *The Skeptic* also had its original stimulus from CSICOP. Wendy Grossman, an American writer living abroad, notes that she is the kind of person "who is excited by unexpected, natural explanations." In recalling how it all started, she says,

> That personality trait made me receptive in late 1986, when Mark Plummer, then the executive director of the Committee for the Scientific Investigation of Claims of the Paranormal [now the Committee for Skeptical Inquiry, or CSI], said to me, "Do you think you could start a newsletter over there?" I was living in Dublin, where I knew hardly anyone, and I had been reading the committee's own publication, *Skeptical Inquirer*, for more than five years after running across first a live lecture/demonstration by magician and debunker James Randi and then a copy of *Science: Good, Bad, and Bogus* by Martin Gardner.[10]

Some things work out in nonobvious ways, she says: "If you ask people now, 22 years later [she was writing in 2009], what *The Skeptic* has changed, you won't necessarily get an encouraging response." Alternative medicine booms; books, magazines, and TV shows on the paranormal proliferate; and people still know their star sign. She continues, "On the other hand, there seem to be a lot more skeptics—and a lot more visible skeptics—all over the landscape, and when you're a founder that seems like a result."[11]

Before it began, she "can remember practically throwing things at the television" over a daytime discussion of spiritualism with virtually no skepticism. She recalls thinking, "Why wasn't there a skeptic to question whether there were any spirits to begin with? That doesn't happen now. . . . Within a couple of years of *The Skeptic*'s founding, you wouldn't see a TV show promoting paranormal claims without a skeptical viewpoint." She continues, "It feels, anyway, as though there are a lot more skeptics and skepticism around . . . than there were in 1986. How much of a role *The Skeptic* has played in that can be debated by others."[12]

In 1996, *The Skeptic* published a notice looking for like-minded people to start a membership organization in the United Kingdom. The Association for Skeptical Enquiry (ASKE) was founded in 1997 and represents Britain as a member of the umbrella organization European Council of Skeptical Organisations (ECSO). The sprightly "Skeptics in the Pub" monthly meeting was "founded in my living room while stuffing magazines into envelopes one day in 1999," Grossman says. "My sole contribution was to say, 'Sounds great. Go for it.'"[13] It was soon attracting standing-room-only crowds every month and was copied around the country.

Grossman, twice editor of *The Skeptic* (she moved from Ireland to Britain early on), shared these reminiscences on the occasion of the magazine's twenty-first anniversary in the introduction to a book-length anthology of forty-one of its articles, *Why Statues Weep: The Best of* The Skeptic.[14] It is a fine representation of the kinds of substantive and always interesting articles *The Skeptic* has published.

The Skeptic ceased publishing a printed edition in 2016 and went totally online (its back issues are all still available) at https://skeptic.org.uk. Energetic skeptic Michael Marshall is now its editor. Former editors in chief Chris French, Deborah Hyde, and Wendy Grossman remain involved as writers, advisory editors, and members of the Editorial Advisory Board. At the request of the British Library, *The Skeptic* website is archived on the UK Web Archiving Consortium's project to "concentrate on sites of cultural, historical, and political importance."[15]

At any rate, together these three major English-language journals of scientific skepticism—the *Skeptical Inquirer* (US), *The Skeptic* (Australia), and *The Skeptic* (UK)—have combined to bring informed critical inquiry to audiences in all hemispheres. Add to them all the other journals in other countries and languages, such as Germany's *Skeptiker*, the Netherlands' *Skepter*, and Latin America's Argentina-based *Pensar* for the Spanish-speaking world (now online and published by CFI), and you can see that a significant portion of the globe is well covered.

CSICOP, Kurtz, Randi, and others helped inspire the formation of skeptic groups virtually everywhere. Longtime CSI executive director Barry Karr remembers all this well. Karr is at the center of everything the US organization does. He told me in 2021,

> We sent Mark Plummer and Wendy Grossman on a tour of Europe to help set up skeptical groups in a variety of countries. We sent out letters to readers inviting them to an event and from those events groups were established. We had Massimo Polidoro and Lorenzo Montalli come to stay with us in Buffalo for many months to see how an organization worked. Randi, Kurtz, Plummer, and I traveled all over Europe, the United States, and other places to meet with skeptics and try and organize them. Plummer was particularly active on this front. We were instrumental in helping to set up the European Council of Skeptical Organizations.

In the 2001 anniversary volume *Skeptical Odysseys*, Kurtz proudly writes of the requests that poured in over the years asking for CSICOP's help in forming new groups: "Today there are approximately one hundred skeptic organizations in thirty-eight countries and a great number of magazines and newsletters published worldwide, and they continue to grow."[16]

In addition to helping form skeptic groups worldwide, CSICOP often participated in major international skeptic conferences. Notable locations include Heidelberg (1998); Sydney, Australia (2000); Moscow (2002); Padua, Italy (2004); Berlin (2006); and Beijing (2008). The World Skeptics Conference in Abano Terme, Italy, in 2004, hosted by the Italian skeptic group Comitato Italiano per il Controllo delle Affermazioni sulle Pseudoscienze (CICAP), took place near Padua, where Galileo taught, wrote *The Starry Messenger*, and discovered the moons of Jupiter. One of its memorable sessions, "The World of Galileo Galilei," featured a moving homage to Galileo's courage. Italian physics professors and science historians talked about his life, discoveries, and conflicts with authorities. Interspersed were clips from a little-known Italian movie about Galileo's battles with church leaders. A physicist brought onstage a large wooden inclined plane and measuring instruments like those Galileo used in his experiments demonstrating that objects fell at the same velocity regardless of their weight. He showed how Galileo did the experiment and said, "The scientific method was the major contribution of Galileo."[17]

Also memorable was a major conference in China. Titled "The World Congress on Scientific Inquiry and Human Well Being," it was a joint endeavor of the CFI and the Chinese Research Institute for Science Popularization (CRISP). Speakers included CSI fellows and Nobel laureate scientists Murray Gell-Mann and Sir Harry Kroto, philosopher Daniel

Dennett, physicist Lawrence Krauss, Paul Kurtz, and China's Lin Zixin. One remarkable moment came when one of China's speakers, Qin Dahe, a noted geographer and climatologist and member of the IPCC working group 1 on the physical sciences basis of climate change, announced to the audience that he had just learned overnight that the IPCC had been named cowinner of the Nobel Peace Prize, making him a cowinner too. The applause was loud.

For its first quarter-century, CSICOP was the only such national group in the United States. In 1992 came Michael Shermer's Skeptics Society and its new quarterly magazine, *Skeptic*. It started just in time to honor on its first cover Isaac Asimov, who died April 6, 1992. This enterprise, too, had its roots in CSICOP's outreach activities, getting its start from solicitations to *Skeptical Inquirer*'s Southern California mailing list to what was then a regional Los Angeles–based group. Since then, there have been *two* major skeptic magazines in the United States (not counting the many newsletters by local groups). As for the United Kingdom, CSICOP was long closely associated with its quarterly magazine *The Skeptic*, publishing it out of Amherst for several years.

Conferences and CSICons

Over its first thirty years, CSICOP held conferences about every year and a half, usually near a major university and in conjunction with such relevant academic departments as physics, psychology, and philosophy. These tended to be very substantive conferences, close to academic in tone but, given the subject matter, dealing with far more controversial and emotional (and interesting!) topics. They were great fun, as well. They gave a chance for science-minded skeptics from all over to get together and hear from one another in person. Then for a time, starting about 2005, that tradition went into a seven-year hiatus. The James Randi Educational Foundation (JREF) began filling that gap with a new series of popular conferences, "The Amazing Meeting" (TAM), held in Las Vegas. Nevertheless, CSICOP continued to sponsor all sorts of smaller skeptic events during that period. These included its popular annual Skeptic's Toolbox workshops in Oregon and two cruises with lecturers (2005 and 2009). It cosponsored or sent delegates to conferences of the ECSO in Spain,

Dublin, Budapest, and Stockholm; a Pensar conference in Latin America (Lima, Peru); and a lecture tour in China in 2005.

CSICOP resumed regular conferences in October 2011 with CSICon 1 in New Orleans and CSICon 2 in Nashville in October 2012. The conference in 2013 was a joint "summit" conference in Tacoma, Washington, with its supporting organization, the CFI, and sister organization, the CSH. Its next was also a joint conference at CSI's founding location, Amherst, New York, on June 11–14, 2015.

Since then, CSICons have all been held annually in Las Vegas, except for a pandemic-caused hiatus in 2020 and 2021. They resumed in October 2022. In its place during those two years, executive director Barry Karr arranged a series of online talks, *Skeptical Inquirer Presents*. Hosted by comedienne Leighann Lord (also cohost of CFI's *Point of Inquiry* podcast), they featured leading experts on conspiracy theories, countering antivaxxers, cognitive dissonance, medical myths, science denial, the infodemic, QAnon, and related topics. "When I am looking for speakers, I now find myself thinking in the back of my mind: I want people to tell me (us) what the heck is going on!" Karr says. "I guess the overall theme we've been dealing with, in a broad sense, is 'Misinformation: Why People Fall for It and How to Combat It.'"[18] The series continued through 2022 and may become permanent.

The amazing entrepreneurial academic Paul Kurtz founded all these organizations. He raised the funds to build CFI's headquarters, across the street from the Amherst campus of the University at Buffalo, where most of CSI's and CSH's activities are still located. (Some editors and employees have always worked elsewhere; in this internet age, having everyone at one central location is no longer essential.)

Agonizing Transition

For about five years, from 2007 to 2011, the combined organization went through an agonizing period as it transitioned to new leadership. CSI had a single charismatic founder everyone loved and respected, but he couldn't go on forever. Kurtz was well into his eighties then. For some years, the boards of the three organizations had worked with him to try to come up with a succession plan. They knew replacing such a remarkable visionary was near impossible. They tried various ideas, but none got far. (I am on those boards. I can tell you the whole process was extraordinarily difficult and painful for everyone involved.)

Things eventually stabilized again. Ronald A. Lindsay, a strong defender of secularism and scientific skepticism, became the president and CEO of the combined organization. He was a calm and steady force. Like Kurtz, Lindsay had a PhD in philosophy, but he also had a law degree, increasingly relevant in these times. The CFI board, chaired by Eddie Tabash, a civil liberties lawyer deeply dedicated to the organization's causes, began taking a much more active role to ensure the longtime vitality of the organization.

When Lindsay retired, the board appointed Robyn Blumner, a former ACLU lawyer, a former newspaper editorial writer, and the director of the Richard Dawkins Foundation for Reason and Science, as CEO and president. She has continued to oversee a further expansion and stabilization of CFI. At virtually the same time (January 26, 2016), the Richard Dawkins Foundation merged with CFI. This created a much stronger and more resourceful organization. Dawkins, the famous evolutionary biologist and author of *The Selfish Gene*, one of the most notable scientific books of all time, became a member of the CFI board. He had long been a fellow of CSI, a prominent advocate for good science and clear thinking, and an opponent of pseudoscience in all its forms. The merged organization became the largest in the United States with a mission of promoting secularism *and* science. All this was announced in the May/June 2016 *Skeptical Inquirer*.[19]

Paul Kurtz, the extraordinarily gifted and pragmatic philosopher and creator of the world's leading organizations for scientific skepticism and secular humanism, died October 20, 2012, two months short of his eighty-seventh birthday. He was my mentor and my friend. I felt privileged to be invited to speak at a memorial service the Kurtz family put on at the University at Buffalo on December 1, 2012. So many people came to deliver tributes that the event lasted the entire day.

Karr and Council

Barry Karr, who first began working for CSICOP fresh out of college in the early 1980s and has always been strongly dedicated to the skeptic mission, remains CSI's executive director. He has long been a steady, stabilizing influence, involved in every aspect of CSI's work and now a valuable repository of its historical memory.

As a young staff member, Karr accompanied our small CSICOP Executive Council delegation to China in 1988 and marveled with amusement at the visage of psychologist Jim Alcock, "who is quite tall, and Randi, who wasn't," drawing a crowd wherever they went. In a 2021 special issue of the *Skeptical Inquirer* filled with tributes to James Randi after his death at the age of ninety-two, Karr poignantly reminisces, "People were amazed at the juxtaposition." He told the story of how once, when Randi was on the *Tonight Show* with Johnny Carson (not the famous appearances with Uri Geller or Peter Popoff but a later one), Randi called him at his parents' house. "We had all watched him on the Carson show, and now he was talking to my mother asking her what she thought of the show, etc. I think it was that moment that my family realized that I was doing important work."[20]

Karr writes regular fund-drive appeals to *Skeptical Inquirer*'s readership. His letters are always chatty and interesting. (One during the pandemic shutdowns in October 2020 included pictures of his dog Dakota walking around the empty offices of CSI and CFI.) He often airs his own anger and frustration at the nonsense swirling about that the organization must help combat. In one letter, he lamented how this late in the pandemic there still were people contending that COVID-19 was a hoax and that wearing masks doesn't save lives. "You would think people would know [better]," he wrote. "But no, we remain mired in false information, conspiracy theory, and magical thinking. We have politicians, religious leaders, and media talking head pundits promoting antiscience, hoaxes, and lies as the truth." He ended with a postscript: "Won't you help us and everyone else get back to pre-COVID-19 normalcy? Dakota wants to come see her friends."[21]

For a brain trust, CSI has a small Executive Council who provides advice and guidance and about 120 prominent CSI fellows, elected by the Executive Council for their distinguished contributions to science and skepticism. In June 2018, the Executive Council formalized its criteria for election as a fellow:

1. Outstanding contribution to a scientific discipline, preferably, though not restricted to, a field related to the skeptical movement

> 2. Outstanding contribution to the communication of science and/ or critical thinking; or
> 3. Outstanding contribution to the skeptical movement.
>
> Fellows serve as ambassadors of science and skepticism and may be consulted on issues related to their areas of expertise by the media or by the Committee.[22]

As a measure of the group's internationalism, by early 2021, fellows heralded from thirteen countries: Australia, Brazil, Canada, China, France, Germany, India, Italy, Mexico, the Netherlands, New Zealand, the United Kingdom, and the United States.

Also, over the years, CSI became a much more gender-diverse organization than when it started. For a long time now, many outstanding women scientific skeptics have been deeply involved as fellows, members of the Executive Council, authors, and speakers at CSI conferences. Among them are parapsychology critic Susan Blackmore, psychology professor and memory expert Elizabeth Loftus, psychologist and author Carol Tavris, anthropologist and evolution educator Eugenie Scott, astronomer and search for extraterrestrial intelligence (SETI) expert Jill Tarter, *Cosmos* cocreator Ann Druyan, philosophers Susan Haack and Barbara Forrest, physician and medical pseudoscience critic Harriet Hall, physician and medical editor Marcia Angell, science historian Naomi Oreskes, evolution teaching educator Bertha Vasquez, and "Guerrilla Skeptic" Susan Gerbic.

CSI Today

Now let's zoom ahead. CSICOP continued, as Kurtz had written after its first five years, "to encourage the spirit of scientific inquiry and to cultivate the skeptical attitude and the quest for evidence and reason."[23] CSI (formerly CSICOP) is now forty-five years old. Over that time, it has attempted to keep aglow the light of reason and rationality and to cultivate scientific thinking in the wider public. It has critically examined thousands of individual claims and assertions and published the results for the world to see. It has explored virtually every issue important to science-minded skeptics. It has encouraged greater skepticism in the news media and served as a source of reliable scientific information for newspapers and television. It has done its best to make others aware of the dangers to a democracy of all confusions between reality and fantasy, sense and nonsense, and real science and its pretenders and adversaries.

The task has never been easy. There have been epic battles and setbacks. CSICOP was battered by internal controversies and public controversies alike. But CSICOP and now CSI have always persevered. Furthermore, the committee has always had the involvement and support of the world scientific and academic communities. Scientists and scholars—and tireless writers and investigators—voluntarily contribute their energy, expertise, and scholarship to critically examine claims and explore all the social, political, scientific, philosophical, and educational issues surrounding them. Good-spirited people everywhere who crave responsible, scientifically evaluated information and perspective continue to look to it for authoritative analyses. Scientific skeptics, no matter where they labor, in cooperation with researchers in dozens of academic fields, help provide that.

The *Skeptical Inquirer* through all these years has continued to tackle the new manifestations of all the old classical topics of pseudoscience and fringe science—everything from UFOs and ghosts to bigfoot and dozens more—while expanding its reach to topics of broad import where science misinformation affects society. It started taking on misinformation about climate change in 2008. It lost some subscribers as a result but gained the respect and support of others. Many *Skeptical Inquirer* authors have sought to explain how our brains work (and don't work) to sometimes misconstrue what we read and see and hear according to our own beliefs and predispositions. And they have explained all the other psychological reasons we embrace ideas and theories that seem pleasing or interesting or fascinating but may or may not have any connection to the real world and why, when we are confronted with evidence conflicting with our most deeply held beliefs, we discard the evidence, not the beliefs. It is just something we humans do—some more than others. As I write in one *Skeptical Inquirer* commentary, "Our quest seeks to understand not only the external world of nature out there, but our own selves, what makes us human—wonderful and creative, flawed and exasperating."[24]

Awash in Disinformation

In 2020, the nation was awash in conspiracy theories, misinformation and disinformation, and outright lies spread effortlessly on social media and partisan cable news outlets. Some of these were about the COVID-19 pandemic then afflicting the world, and some were a result of purposeful political action—including from Russian bots seeking to undermine our sense of our own democracy; fear-mongering extremist hate groups seeking to spread their views to the public; and a president who cared little for the truth and instead fostered lies, misinformation, and conspiracy theories that furthered his political interests. The *Skeptical Inquirer*, like most responsible news outlets, including other science-oriented magazines, did its best to report on these issues and examine their consequences. For a time, every issue of *Skeptical Inquirer* was virtually filled with news and articles about the pandemic misinformation. CFI set up an online Coronavirus Resource Center compiling rebuttals of misinformation. Then when Trump lost his bid for reelection in November 2020 but insisted—vociferously, repeatedly, in contradiction to all evidence—that he won, the focus on misinformation and lies necessarily turned to him and the political system that embraced him and created and nurtured that misinformation. Things became even worse when his words helped foment the January 6, 2021, assault on the Capitol as Congress was attempting to carry out its constitutional duty to certify the election.

This situation was uncomfortable for a nonpartisan publication and organization whose mission is critical thinking and evidence-based analysis. Some partisans, in letters the *Skeptical Inquirer* published, criticized the magazine for becoming "political." In a response, I, as editor, disputed that complaint: "We at SI have spent four-plus decades excoriating pseudoscience (and antiscience) from every source imaginable, even recently from some Nobel laureate scientists, but when we point out that it was coming from our president, that is 'politics.'" It doesn't work that way. Politicians don't get a free pass to spread nonsense about scientific information without rebuttal. I could do no better than echo the words of the editor of *Science* magazine, H. Holden Thorp, who had received similar complaints urging his distinguished magazine to "stick to the science." Said Thorp, "We are sticking to the science, but more importantly we are sticking up for science."[25]

The *Skeptical Inquirer* continued to stick up for the science, as well, and published still more news, feature articles, columns, and reviews about conspiracy theories (QAnon got a two-part examination) and intentional misinformation and lies. But, as I said in an interview, I ardently looked forward to the time, hopefully very soon, when we could return mainly to classic topics of pseudoscience.

Internet and Social Media

The need for informed scientific skepticism has never waned, but the circumstances in which it operates have greatly changed. The internet has led to the proliferation of pseudoscientific and antiscientific claims, conspiracy theories, and all other forms of misinformation. This has made skeptics' jobs much more difficult. Sometimes the misinformation and disinformation seems to swamp all efforts to rebut it, but the internet also created an opportunity to reach wider audiences quickly with good science and informed responses. (An example is *Skeptical Inquirer*'s Facebook page. Managed by Karr, it is a wonderful, lively way to keep up with daily developments, not just in scientific skepticism, but in science, as well.)

What once was a fairly homogeneous, organized, and somewhat centralized set of skeptical activities has now become a much more diverse, prolific, and fragmented situation. Young people and women can easily enter and create their own niches, and that is good for everyone. Many disparate groups have sprung up. Some have no connection with each other or any sense of the history of this movement that I recount here. Countless investigators and skeptics of all stripes now have their own blogs, podcasts, and online columns. Whole new forums have been created, such as the popular *Skeptics Guide to the Universe* podcast by physician/skeptic Steven Novella (another noted CSI fellow) and his brothers and colleagues and CFI's own podcast of in-depth interviews, *Point of Inquiry*. It is getting harder and harder to keep up, but for the most part, all this is a healthy proliferation of science-based and evidence-based skepticism.

Opinion is always easy to come by, but hard-won facts and evidence; historical and cultural perspective; and a mature, scientific viewpoint are still most important. That requires considerable discipline and work. The *Skeptical Inquirer* has always tried to live up to high standards of inquiry: state things carefully, let the facts speak for themselves, provide balanced analysis, give thorough documentation, avoid personal attacks. At CSI and in *Skeptical Inquirer*, as I've said, we are at least a *semi*-scholarly journal, and we do professional scientific journalism: strongly investigative, hard-nosed to be sure, but still professional. In this age of personal blogs and posts, it sometimes seems that these kinds of standards have gone out the window. I think they remain essential.

What CSI and other skeptic organizations worldwide are really defending and advocating are certain crucial high values. We are ultimately defending a variety of liberties essential to enlightened societies and functional democracies: freedom of inquiry and expression at all levels; science as the most reliable guide to truth about the natural world; a love of learning and questioning; respect for human rights and dignity; the right to learn, to study, and to investigate unfettered by any authoritarian interference; free and open and civil discussion of all issues; and the idea that hard-won new knowledge—while it always raises new issues and problems—is essential for continuing human progress.

CHAPTER TWELVE
FINAL THOUGHTS, FUTURE HOPES

This book starts with a look at a number of recent scientific discoveries that piqued my interest and I hope yours, as well. How many will stand the test of time? Perhaps most? Perhaps just some? Science at its frontiers is filled with uncertainties. It is high risk, high gain. This is what makes it exciting. As chapter 7 shows, this acceptance of uncertainty is one of the least-appreciated values of science and one of its most startling contrasts with pseudoscience, where uncertainty is anathema. To begin this final chapter, let's look at a couple more major scientific and technological achievements. One is already familiar to almost everyone; the other, likely not.

The James Webb Space Telescope

Our understanding of the universe always takes vast leaps forward when we acquire new and more powerful observational techniques. The Hubble Space Telescope was one of those, and now the James Webb Space Telescope has brought our observational abilities into a new realm. It observes the universe from far out in space (a million miles from Earth) and differs from Hubble in two other main ways: Its primary mirror—and therefore its collecting power—is much larger, and it observes the universe in infrared light, while Hubble (still going strong) observes primarily in visible and ultraviolet light.

Because the earliest and therefore oldest and most distant galaxies have had their light stretched or red-shifted over the eons as the universe expanded, the light from them that has reached us is primarily in the infrared part of the spectrum. This makes the James Webb Space Telescope a powerful tool for observing those very early galaxies as seen near the beginning of the universe 13.5 billion years ago. The first scientific image released from the James Webb Telescope, in July 2022, is a "far-field view," stretching almost that far back in time. It shows a region of sky the size of a grain of sand held at arm's length, just a tiny, tiny part of the whole sky. Yet that incredibly small area of sky is filled with distant galaxies nearly 13 billion years old, seen as they were in their infancy.

Infrared has another advantage, as well. Visible light is easily blocked by clouds of dust and gas (that is why the center of our galaxy is so difficult to see), but infrared light can penetrate dust and gas, so we can see areas of star formation and regions near galactic centers that can't be seen any other way.

It is not the first orbiting space telescope to observe in the infrared. The Spitzer Space Telescope, launched in 2003, did the same, but Webb's mirrors are sixty times larger than Spitzer's and seven times larger than Hubble's. And that was one of the challenges. How do you make a mirror 21 feet (6.5 meters) in diameter and launch it into space? No mirror that large had ever been launched into space before. The answer was to build the mirror in eighteen separate segments, each hexagonal in shape and 4.3 feet (1.32 meters) wide, and then fold them like origami into the shell of the Ariani 5 spacecraft that would carry it into orbit. Each mirror is a precision device in itself, made of lightweight beryllium and coated with a seven-hundred-atom-thick layer of gold. Of course, this means that the mirrors would have to unfold in space, a complicated and intricate ballet, all on their own and without help from Earth. I remember the delight when I learned the unfolding process was proceeding properly. When the final mirror segment moved into place, the telescope mirror system was complete. To achieve perfect focus, 132 tiny motors, or actuators, adjust parts of each mirror segment so that the effect is one single, large mirror. This was all happening while the telescope was being rocketed toward its eventual permanent orbital position a million miles from Earth. This is really out there—nearly four times farther away than the Moon.

Just think of all the things that could go wrong in just this one part of the setup process. NASA's scientists, engineers, and administrators were fully aware of the risks, but they had confidence because of the rigorous testing all the procedures had undergone on Earth. But this was space, and not every condition can be perfectly modeled in laboratory settings on Earth.

You can understand the exuberant shouts of joy when word came that the telescope mirrors had properly moved into place and then again later, when the first scientific images showed all was working perfectly. A family member of mine is an accomplished optical physicist. He spent his career at a national laboratory designing optics for satellites. He confided to me after the first Webb images were released, he had feared the space telescope was not going to work; there were just too many complications. He was very happy and surprised to be shown wrong.

The James Webb Space Telescope is a marvelous human achievement. It reminds us that we human beings can indeed do great things when we put our minds to a task and harness our resources in imaginative ways to create something new that will allow us to make discoveries in the future that we can't even anticipate right now. This is the scientific spirit of adventure and discovery. It is inspiring and shows the interaction of technology and science. One feeds the other in a complex process where new advances in applied science and basic science build on each other; together, they can achieve new capabilities and new discoveries.

And it again demonstrates the collaborative nature of most good science these days. This is another poorly appreciated value of pseudoscience. Webb is an international collaboration among NASA, the European Space Agency (ESA), and the Canadian Space Agency. ESA launched the rocket carrying Webb from its launch facility in Kourou, French Guiana, on Christmas Day 2021 and provided the instrument that detects near-infrared light. Canada provided a fine-guidance sensor and one of the spectrographs that will study the light from stars and the atmospheres of exoplanets. It truly is a telescope for the world. Thousands of scientists, engineers, and technicians from fourteen countries helped design, build, test, and integrate it. Scientists from forty-one countries were awarded observing time for the first year of Webb's scientific observations.

The Webb telescope will look into the distant past of our cosmos. It will study the atmospheres of some of the thousands of exoplanets we have discovered in recent years. It will look through disks of dust around stars that could be more planets in the making. It will do all those things and more.

Astronomers got a better taste of what the Webb telescope could do just a few weeks after the first photos were released. It was a big surprise: From its first images back in time, they could see tens, hundreds, maybe even a thousand times more bright galaxies in the early universe than they had expected. These early galaxies were forming stars at a furious rate. An article in *Science* quoted one of the astronomers from UCLA saying his team were seeing these early galaxies "form stars like crazy." He said they look "like giant balls of star formation and nothing else."[1] So that all was a great start, but it is what we can't quite anticipate that sets us up for wonder and reshaping our thinking. We can all look forward to whatever new discoveries this novel instrument of human progress has given humanity.

Solving the Protein-Folding Problem

Okay, I agree at first glance, this advance may not sound particularly exciting or important. My scientific colleagues assure me it is both. They say it will revolutionize all areas of molecular biology, structural biology, and computational biology, and it has profound consequences for understanding how proteins interact with other molecules.

Proteins are the workhorses of our cells. They contract our muscles, convert food into energy, transport oxygen in our blood, fight microbial invaders, and do myriad other jobs. Each protein is a linear chain of up to twenty different amino acids, which is all encoded in our DNA. These long chains, different for each protein, have to fit inside the cell, and they do so by folding up in exquisite, three-dimensional shapes. Each protein's shape is different. The first two protein structures were determined in 1957 by John C. Kendrew and Max F. Perutz.

An American biochemist, Christian Anfinsen, and others first suggested that it is interactions between amino acids that pull proteins into their final shapes, but the number of possible combinations is astronomical: In the late 1960s, one American molecular biologist calculated that for a protein chain to cycle through all the possibilities one by one, it would take longer than the age of the universe. Of course, that's not how nature works. In fact, in a mere instant, each protein folds up into one distinctive shape. The precise shape, we now know, determines how each protein works and what it does.[2]

In his Nobel Prize acceptance speech in 1972, Anfinsen expressed a hopeful vision: Someday it would be possible to predict the 3D shape of any protein just from its sequence of amino acid building blocks. Scientists could already detect their shapes using X-rays to bounce off their atoms, a process called X-ray crystallography (Kendrew and Perutz's 1957 technique), but it was phenomenally time consuming and expensive. Today, 185,000 structures determined this way are in the Protein Data Bank. But that just scratches the surface.

In the last few years, an entirely new way of determining this vast array of complex 3D shapes has become possible. It involves artificial intelligence (AI)–driven software that can churn out accurate 3D protein structures by the thousands. This is far faster than the computer models researchers first developed in the 1990s to do this task. One thing that made the new achievements possible stems from a unique competition set up in the 1990s to find better and faster ways to predict proteins' 3D structures. This stimulated a lot of research.

Over the years, the models got better and better, but around 2018 came a quantum leap in predictive capability. A program called AlphaFold, an AI-driven program developed by Google sister company DeepMind, trains itself on databases of known three-dimensional protein structures to learn how to predict the shapes of unknown ones. Its successor in 2020, AlphaFold2, did even better, achieving a score on par with the experimental techniques but of course acting much, much faster. In 2021, things accelerated further. Another AI program, RoseTTA-Fold, developed by University of Washington computational biologist David Baker, solved hundreds of protein structures, all of them important to the drug industry. A week later, DeepMind scientists reported they had done the same for 350,000 proteins found in the human body, about 44 percent of all human proteins.[3] By late summer of 2022, DeepMind said it had now deciphered the structure of almost every protein known to science. It was an astounding achievement. A European molecular biologist called the new protein database a "gift to humanity."[4]

Like determining the structure of DNA (Francis Crick, James Watson, Rosalind Franklin, and Maurice Wilkins in 1957) or determining the structure of the human genome with its sequences of three billion chemical base pairs that make up human DNA (announced in 2003 but not fully completed until 2022) or implementing the gene-editing technique called CRISPR (which really got started in 2007), these kinds of advances can have profound effects that we can't initially comprehend.[5] Scientists are convinced the new protein-folding-prediction methods will forever change biology and medicine.

This advance in predicting the 3D structure of proteins was so significant that *Science*, one of the world's leading scientific journals, designated it the 2021 Breakthrough of the Year. The journal *Nature* declared it the 2021 Method of the Year. "I thought the protein-folding problem would never be solved," H. Holden Thorp, *Science*'s editor in chief, writes in an accompanying editorial. "It is breakthrough on two fronts. It solves a scientific problem that had been on the to-do list for 50 years. And just like Fermat's Last Theorem or gravitational waves, scientists kept at it until it was done." Also, he says it is a "game-changing" technique that, like CRISPR and cryo-electron microscopy (cryo-EM), "will greatly accelerate scientific discovery."[6]

There is something else about these new advances that is a mark of excellent science. First, the two different AI methods of predicting protein 3D structures, AlphaFold2 and RoseTTAFold, were published online simultaneously on July 15, 2021, by *Science* and *Nature*, respectively, and in their weekly print issues of August 20 and August 26, respectively. This says something about the cooperative nature of science, which I describe earlier as an important value. In the same way, the earlier determination of the human genome was carried out by two entirely different methods and research groups but announced simultaneously in April 2003. (One recalls the simultaneous announcement of Charles Darwin's and Alfred Russel Wallace's epochal discovery of evolution by natural selection, each achieved independently but announced together to the Linnean Society in London on July 1, 1858.)

More important, the two new protein-folding-prediction methods are now free and publicly available. Scientists everywhere can now use them to quickly advance the pace of discovery, which is already happening.[7] This openness—the ready availability of all new data—is another mark of all good science.

The James Webb Space Telescope and these new AI methods for predicting protein 3D structures are just two advances of many in science. They are shining examples of science at the frontiers of discovery. Hundreds, thousands, perhaps tens of thousands of other findings likewise promise to change the course of science and human history. Science's role in human affairs is dynamic, ever changing, ever advancing. It is a generator of change and progress, and it provides a real sense of possibility for better human futures.

Pseudoscience in Contrast

What have we, in contrast, from pseudoscience? I think we need a long pause here.

I'm thinking, I'm thinking.

If we would need thousands of pages to properly chronicle the pace of scientific discovery in any recent time period, then how much would we need to do the same for pseudoscience? Perhaps, perhaps a . . . blank page?

Pseudoscience achieves no gains in scientific knowledge because it is incapable of doing so. It doesn't use the methods of science. It doesn't have science's self-correcting and self-criticizing processes. It doesn't have any of the collaborative nature of science. It doesn't even do any scientific research. Oh, it makes sensational claims, and it develops and markets products, all right. It does do that. It is very good at that. But they are seldom based on any real science.

Products and remedies derived from pretend science have little chance of working in the way promised. Claimed discoveries based on pretend science and human wish fulfillment may cause momentary excitement among fans and the gullible, but soon they are mostly forgotten (by all but the most fervent believers) when they can't be repeated or proved real in any satisfactory way. But that seldom deters. The fans of pseudoscience are always in the quest of the next big thing. The fact that last year's big thing and last decade's big thing and last century's big thing fell quietly by the wayside doesn't seem to detract from their enthusiasm. In some ways, you almost have to admire their trust in their convictions that someday some of it will be shown true and become accepted by all. It is a pleasant fantasy world they live in.

Scientific discovery truly shapes our world—the real world. Historians and political commentors in their writings and analyses usually give precedence to wars, political leaders, and political controversies and decisions as the prime influences on history. But the best among them know that science and technology alter the world in at least equally profound ways. Few, however, fully acknowledge that fact. Notable exceptions: Walter Isaacson and, before him, Daniel J. Boorstin.[8]

Is the World Worse?

None of this means the mystery mongers, conspiracy theorists, fantasy fomenters, and all the other manufacturers of misinformation don't cause enormous damage and harm. This book chronicles a lot of it. I show some of the dynamics and contrast their methods and their attitudes with that of real science. All those processes of misconception and misinformation are still very active.

In recent years, they seem to have taken an even worse turn. The nation and world have been through a frightfully chaotic period. First, we had a global pandemic that shut us all down and caused immense human suffering—while newly developed life-saving vaccines were viewed incongruously by a significant and vocal minority as something to distrust and resist. Then, in the United States, we experienced political upheavals (including an unprecedented violent attack on the US Capitol) fomented by an authoritarian outgoing president spewing conspiracy theories and false claims of election fraud and a "stolen election" and carried out by vocal and violent partisans who believed him. (All this was later documented by the congressional committee investigating the January 6 attack.[9]) It was something that most Americans thought could never happen in any enlightened democracy.

(I am aware that these hyperpartisans still contend that the lies are true and that the election was fraudulent. They remain a powerful force of mass deception and political influence. Ironically, many election deniers among them since have been busy getting themselves elected to state legislatures and other state and local offices around the country, another scary trend.)

Russia invaded Ukraine in a brutal ground war of relentless bombardment reminiscent of World War II. Gun violence, murders, and mass murders in the United States soared. It has all been jarring, disheartening, disillusioning. Here we are in the first third of the twenty-first century, and Edvard Munch's 1893 painting *The Scream* could be the portrait of our time.

Equally, all manner of other conspiracy theories and out-and-out pseudoscientific beliefs reverberate throughout the nation and the world. The antiscience of the recent antivaccine movement has achieved an almost cult status, fired by misinformation and a newfound distrust of medical science. The anti-climate-science movement may have changed tactics over the decades, but it, too, remains a powerful counterforce that has hindered our doing much about that global threat—while more and more extreme weather events pummel regions around the world.

The Longer Perspective

So what do we do? Is the world collapsing around us? Have all common sense and rational behavior left us? Is there no reasonable response other than despair? It sometimes seems so, but we should have heart. A positive, optimistic attitude is one part of that. It helps shape how you see the world, but we needn't rely on just our own human psychology to see better things ahead. We can rely on the data. Like all good scientific thinking, we can see what solid data, real, fact-based information, tell us about global trends that affect human welfare. And by virtually every standard, things are improving worldwide—and they have been getting better for quite a long while.

If this seems counterintuitive to you, you are not alone. The day's news is usually abysmal. Television news and social media inform us immediately of all that's bad in the world. We cannot escape it. We hear all of that because these kinds of events are newsworthy, dramatic, unusual. Human tragedies told in emotional news stories are compelling. They remain seared into our memory. But the hallmark of rational thinking is perspective. When things go well, few take note. When a storm fades and floodwaters recede, a major political crisis wanes, or the airlines go for months without a serious crash, few news organizations report it. When good things happen, that is seldom news. Life goes on all the time, more or less the same or even a little better, but that's not what we hear.

There are two books I think everyone should read. Remarkably, both were published in 2018:

1. *Factfulness* by Hans Rosling and colleagues
2. *Enlightenment Now* by Steven Pinker[10]

Any journalist, political commentator of whatever stripe, politician, economist, business leader, social scientist, of informed citizen—all of us concerned about the state of our world who profess the need to understand it as it really is—will find their views challenged in necessary ways by these two books. And while they both are well argued, their conclusions are similar, and those don't depend on argument—not at all. They depend on the data. Yes, cold hard data, something many shy away from.

Hans Rosling's book is the product of a remarkable program of research carried out by the Gapminder Foundation, which he and his colleagues founded. The book's subtitle tells all: *Ten Reasons We're Wrong about the World—And Why Things Are Better Than You Think.* Rosling became a physician and then a professor of international health. He advised the World Health Organization and UNICEF. He soon became concerned about people's misconceptions about the world and buried himself in statistical data, becoming a passionate statistician, a chronicler of the health of the world. But he brings statistics to life. He has given wildly popular TED talks and refers to his "lifelong mission to fight devastating ignorance. Previously I armed myself with huge data sets, eye-opening software, an energetic lecturing style, and a Swedish bayonet for sword swallowing. [He tells that anecdote early in the book.] It wasn't enough. But I hope this book will be."[11]

I was extremely fortunate to see Rosling and his son Ola Rosling and daughter-in-law Anna Rossling Rönnlund, his collaborators and coauthors, in action once on a stage in Stockholm, Sweden, long before their book was published. The occasion was the Fifteenth European Skeptics Congress, organized by the Swedish Skeptics and the European Council of Skeptical Organizations (ECSO), on August 23–25, 2013. (I spoke on the final day.) They led off the conference.

Their presentation that first morning was remarkable. It was a rapid-fire "fact-based worldview with animated data." At the beginning, they gave everyone in the audience an electronic feedback device. On it, we could record our responses to their questions. I'd never seen a conference start this way. They then offered a series of questions about the state of the world: "In the last 20 years, the proportion of the world population living in extreme poverty has . . ." —we could answer "almost doubled," "remained more of less the same," or "almost halved"—and so on through a broad litany of questions about global trends in human welfare. We punched in our answers. Software they had developed called Trendalyzer instantly compared our answers with a vast array of global statistics. We were, we thought, a cosmopolitan audience. Intelligent. Worldly. Well informed.[12]

Well, not so. We were wrong. We got most of the answers wrong. Almost everyone in this multinational audience got the majority of answers wrong. All this was immediately displayed on screens onstage. It was eye-opening, humiliating, chastening. Much of what we think we know about world population trends is wrong. A dramatic new reality has set in, Rosling told us. For instance, on population: Whereas couples used to have an average of six children, of whom four died, the new balance is now two kids, with most of them surviving. This dramatic decline in babies born per woman is true of all religions and all regions. Europe just started doing it earlier. As a result, they assured us, world population will continue rising but not at the steep rate predicted four decades earlier. And so on.[13]

In their book, the Roslings start by presenting thirteen such questions. In 2017, they had asked 12,000 people in fourteen countries to answer their questions about the state of the world. The respondents scored on average just two correct answers out of the first thirteen. No one got them all right. A stunning 15 percent scored zero![14] Rosling is very blunt:

> Perhaps you think that better-educated people would do better? Or people who are more interested in the issues? I certainly thought that once. I have tested audiences from all around the world and from all walks of life: medical students, teachers, university lecturers, eminent scientists, investment bankers, executives in multinational companies, journalists, activists, and even senior political decision makers. These are highly educated people who take an interest in the world. But most of them—a stunning *majority* of them—get the answers wrong. . . . It is

> not a question of intelligence. Everyone seems to get the world devastatingly wrong.[15]

But it is worse even than that, for the wrong answers are all in one direction: "Not only devastatingly wrong, but *systematically* wrong. By which I mean that these test results are not random. They are worse than random: they are worse than the results I would get if the people answering my questions had no knowledge at all."[16] Monkeys could do better just by randomly pushing the buttons, he laments. They'd at least get 33 percent right (there were three choices of answers per question).

Worse still, the human errors all trended in one direction: "Every group of people I ask thinks the world is more frightening, more violent, more hopeless—in short, more dramatic—than it really is." Rosling continues, "Step by step, year by year, the world is improving." Not everywhere, not all the time. There are always downturns and bad things happening, but as a rule, "though the world faces huge challenges, we have made tremendous progress. That is the fact-based worldview."[17]

Most people don't believe that. It is counterintuitive and contrary to what we have been led to believe. Rosling is well aware; he often heard people saying so even after leaving his lectures, where they had just seen the data supporting his conclusions. We have been led astray by our own brains to embrace an overly dramatic worldview. It is the way our brains work. Mostly we do our thinking quickly, not analytically and slowly. (Economist/psychologist Daniel Kahneman won the Nobel Prize in economic sciences for his research with colleague Amos Tversky demonstrating that in some detail, as he elucidates so well in his book *Thinking, Fast and Slow*.[18]) Most of the time, that works well. It worked especially well in previous times, thousands of years ago. But we live in a very different world today, while our brains are still back there watching for lions and leopards. "Our quick-thinking brains and cravings for drama—our dramatic instincts—are causing misconceptions and an overdramatic worldview," says Rosling. "We need to control our drama intake. Uncontrolled, our appetite for the dramatic goes too far, prevents us from seeing the world as it is, and leads us terribly wrong."[19]

Steven Pinker, the noted Harvard cognitive psychology professor and author, independently came to the same conclusions as the Roslings. In his massive book *Enlightenment Now: The Case for Reason, Science,*

Humanism, and Progress, he amasses reams of global data showing the world is slowly getting better.[20] He includes entire chapters documenting that fact in the areas of life, health, sustenance, wealth, inequality, the environment, peace, safety, terrorism, democracy, equal rights, knowledge, quality of life, happiness, and existential threats.

He shows that data with graphs of improvements in life expectancy, child mortality, maternal mortality, undernourishment, famine deaths, world gross product, GDP per capita 1600–2015, world income distribution, extreme poverty, international inequality, social spending, income gains, US poverty, population and population growth, sustainability, pollution and energy, deaths in all ways (genocide, homicide, cars, pedestrians, plane crashes, natural disasters, lightning, terrorism, etc.), executions, racist manifestations, hate crimes, victimization of children, literacy, basic education, female literacy, IQ gains, global well-being, work hours, retirement, costs of necessities, nuclear weapons, loneliness, suicides, and happiness and excitement—and much else.

Pinker's data show "newborns who will live more than eight decades, markets overflowing with food, clean water that appears with the flick of a finger and waste that disappears with another, pills that erase a painful infection, sons who are not sent off to war . . . the world's knowledge and culture available in a shirt pocket." He attributes these gains, these vast improvements in human flourishing, to the "ideals of reason, science, humanism, and progress." But they are human accomplishments, not birthrights, Pinker notes.[21]

To him, that seems obvious—that the ideals of democracy and human possibility and the fruits of open science and the innovations that discoveries spawn are the ultimate cause of these vast improvements. But he realizes they all need a spirited defense. The ideals of the Enlightenment have always had to struggle with "other strands of human nature," which include "loyalty to tribe, deference to authority, magical thinking, and the blaming of misinformation on evil-doers. . . . Harder to find is a positive vision that sees the world's problems against a background of progress." Furthermore, we have all seen that some countries can slide back into pre-Enlightenment primitive conditions, "and so we ignore the achievement of the Enlightenment at our peril."[22]

The idea of progress, it would seem, should be valued, but Pinker finds it isn't. And he doesn't blame poorly educated people or rednecks or ignoramuses. He blames his fellow intellectuals. He begins chapter 3, "Progressophobia," provocatively: "Intellectuals hate progress. Intellectuals who call themselves 'progressives' *really* hate progress."[23] He doesn't mean that they hate the fruits of progress; they are fine with those. We all love our cell phones and our digital music and our access to medical attention and our airline travel.

When I first read his chapter in the proofs that his publisher sent me, I found it so interesting and revealing that I asked Pinker and his publisher to let us publish an adapted version of it as an article in the *Skeptical Inquirer*. We made it the cover story, "Progressophobia: Why Things Are Better Than You Think They Are."[24] Pinker argues that there is a long intellectual tradition of pessimism and that the idea that the world can get better long ago fell out of fashion. It is the very *idea* of progress that many influential thinkers (he calls them the "chattering class") hate. And that attitude is a powerful resistance, if not to achieving progress, then to realizing we *have* achieved progress, which may be almost as important.

One graph he shows is telling: "The Tone of the News, 1945–2020." Data scientist Kalev Leetaru applied a technique called sentiment mining to every article published in the *New York Times* from 1945 to 2005. He did the same for an archive of translated articles and broadcasts between 1979 and 2010. Sentiment mining looks at emotional words in news articles and categorizes them into positive and negative connotations. The short-term wiggles of ups and downs show the crises of the day, but the long-term trend was downward—steadily downward. In other words, the impression that the news has become more negative over time is real. And it's not just the *New York Times*. News outlets in the rest of the world also became gloomier and gloomier from the late 1970s to the present.[25]

Yet, as Pinker notes, "The world has made spectacular progress in every single measure of human well-being. Here is a second shocker: Almost no one knows about it." It is not about being optimistic. It is about seeing the world as it actually is, a hallmark of a rational person. Pinker concludes, "Learning about human progress is not an exercise in optimism, cheeriness, or looking on the bright side. It's a matter of accuracy, of understanding the world as it really is."[26]

Pinker's book got a lot of praise, but it also got some strong pushback, much of it, predictably, from the kinds of progressives he had already called out. The *New Yorker* gave it a long, thoughtful, and mostly positive review, also referring briefly near the end to the Roslings' book *Factfulness*, but this eminent weekly journal of the literati couldn't be caught ending on an optimistic note, no: "A world in which no one complained—in which we only celebrated how good we have it—would be a world that never improved. The spirit of progress is also the spirit of discontent."[27] Of course, Pinker is not talking about not complaining or only celebrating how good we have it. He is showing us the data that proves, empirically and despite our intuitions, that the world has improved mightily over time and continues to do so.

One year after publication of *Enlightenment Now*, Pinker published on the website *Quillette* a lengthy essay (forty-nine-minute read online) reflecting on the reaction to it and *pithily* responding to some of the most egregious criticisms.[28] He notes that critics insisted that human progress must only be an illusion of cherry-picked data. They even attacked the ideas of the Enlightenment itself, revealing an ill-disguised contempt for science and reason and the idea of human potential. It is hard to believe those are not necessarily values all opinion leaders share, but that is the case. There is too much in this essay to get into here in detail (I highly recommend reading it yourself), but the criticism of cherry-picking the data deserves a firm response, and Pinker is well up to the task: "In each case I chose the most objective and agreed-upon measures. . . . I stuck with public datasets compiled by university researchers, government agencies, and intergovernmental institutions." He avoided numbers from advocacy groups, which could be biased.[29]

He graphed the results: "In every case progress can be seen with the naked eye." The data are there for everyone to see. In contrast, he points out, "journalism, almost by definition, is cherry-picking. It reports rare events such as wars, epidemics, and disasters, not everyday states of affairs such as peace, health, and safety."[30] As a journalist myself who worked for a wire service and newspapers early in my career before turning to science, that criticism hurts a bit, but I know it is true. I realized that at the time. It is not bias per se; it is just that that is the journalist's job to cover the news.

Just one more point about numbers: Scientific thinking is necessarily quantitative. That's not usually the strong suit of liberal arts people, who tend to populate the institutions of politics, journalism, and public affairs. Their innumeracy is a blinder. One line of criticism is that numbers may show one thing, but human suffering isn't really captured by numbers. Pinker responds,

> The accusation is inside-out: looking at numbers is the moral, comprehensive, sensitive way to deal with human suffering. It treats every life as equally precious, instead of privileging members of the tribe or victims that are photogenic or conveniently nearby. . . . Data show us where the suffering is greatest, help identify the measures that will reduce it, and reassure us that implementing those measures is not a waste of time.[31]

Hans Rosling and Steven Pinker (and many others since, including my close colleagues Robyn Blumner and Ben Radford) are trying to tell us something important about the world.[32] First, don't trust our intuition. That's a message that resounds throughout this book. Self-deception is the first deception. Intuitive thinking has its place, but when trying to understand important global trends, the evidence—in this case, objective empirical data—should be the guide. Second, the data show that things really are better than we think. That is a hopeful insight. Why not embrace it?

Conclusion

This is a book about both science and pseudoscience. One of them provides thousands of new discoveries and insights into the nature of the universe and life that inspire us. Some of these discoveries give us a sense of wonder and awe—a feeling of something larger and more wonderful and mysterious outside our own daily concerns. It also leads to innovations and technologies that have the capacity to improve daily lives around the world. These innovations help us to live longer and with a higher quality of life, to stay in communication with friends and loved ones everywhere, and to more quickly realize the potential inside us all.

The other is also interesting and beguiling, but its allure is an illusion. It misleads. It misdirects. It thrives on wishful thinking. It provides false information and false hopes. It is good at disguising itself. Few consumers realize it is bogus. It works in science's shadows and borrows some of science's reflected glory. It exists along a spectrum. Some is fluff and mostly innocent; much is deeply misleading and dangerous. Pseudoscientific remedies and products fleece innocent consumers of their money while marketing themselves as something special and real. Conspiracy theories based on misinformation and lies appeal to our basest fears and are easily manipulated and spread by those with ideological and political agendas. When used to undermine our institutions of democracy, they threaten the very foundations of freedom and progress.

Science is open, merit based, and antiauthoritarian. It thrives best in open societies that treasure human freedom and knowledge and learning. History tells us that magical thinking and pseudoscientific ideas have always been with us and will continue to be. The same is true for conspiracy rumors and conspiratorial thinking. But sometimes the machines of misinformation rise to threaten our accurate understanding of the world to such a degree that we must step up our efforts to counter it. Science is still strongly supported, in open democracies especially, and will continue to thrive where its values are cherished and upheld. But even many scientific leaders, appalled by the virulent opposition to life-saving vaccines during the COVID-19 pandemic and the exponential spread of misinformation and extraordinary claims that have recently afflicted our political systems, have sounded the clarion bell to respond and limit the harm.

We have the capacity to create the kind of world we want. I envision one in which the ideals of democracy resound to the benefit of all. Where learning, knowledge, and scientific discovery are supported and cherished. Where pseudoscience and misinformation, in all their forms, are quickly identified and countered. Where the impulses of antiscience gain no traction. Where, when the world is improving (as the data show), we can appreciate that fact; and where in those times and places where it is not, we can do what's needed to make it better. Where we can all realize our human potential to live fulfilling lives with dignity. Where we all have a sense of possibility and hope for a better future. That is what I would like to leave to the generations that follow.

NOTES

Introduction

1. Isaac Asimov and Jason A. Shulman, eds., *Isaac Asimov's Book of Science and Nature Quotations* (New York: Weidenfeld and Nicolson, 1986), 294.

Chapter 1

1. N. Mangold, et al., "Perseverance Rover Reveals an Ancient Delta-Lake System and Flood Deposits at Jezero Crater, Mars," *Science* 374, no. 6568 (October 7, 2021): 717.
2. Quoted in Ron Cowen, "Gravitational Waves Discovery Now Officially Dead," *Nature* (January 30, 2015), https://doi.org/10.1038/nature.2015.16830.
3. John Hawks, "Comment on 'A Global Environmental Crisis 42,000 Years Ago," *Science* 374, no. 6570 (November 18, 2021), https://doi.org/10.1126/science.abh1878.
4. Andrea Picin, et al., "Comment on 'A Global Environmental Crisis 42,000 Years Ago," *Science* 374, no. 6570 (November 18, 2021), https://doi.org/10.1126%2Fscience.abi8330.
5. Alan Cooper, et al., "Response to Comment on 'A Global Environmental Crisis 42,000 Years Ago,'" *Science* 374, no. 6570 (November 18, 2021), https://doi.org/10.1126/science.abh3655.
6. Alan Cooper, et al., "Response to Comment on 'A Global Environmental Crisis 42,000 Years Ago,'" *Science* 374, no. 6570 (November 18, 2021), https://doi.org/10.1126/science.abi9756.

7. Alan Cooper, et al., "A Global Environmental Crisis 42,000 Years Ago," *Science* 371, no. 6531 (February 19, 2021): 817, https://doi.org/10.1126/science.abb8677.

Chapter 2

1. National Aeronautics and Space Administration, JPL-Caltech, and Space Science Institute, "The Day the Earth Smiled," November 12, 2013, https://saturn.jpl.nasa.gov/resources/5916/; National Aeronautics and Space Administration, JPL-Caltech, and Space Science Institute, "One Special Day in the Life of Planet Earth," July 22, 2013, https://saturn.jpl.nasa.gov/resources/5864/.

2. National Aeronautics and Space Administration, Goddard, and Arizona State University, "NASA Releases New High-Resolution Earthrise Image," December 18, 2015, http://www.nasa.gov/image-feature/goddard/lro-earthrise-2015.

Chapter 3

1. Quantum Analyzer, "What Is Quantum Analyzer AH-Q8? Quantum Resonance Magnetic Analyzer," 2018, http://www.quantum-resonance-magnetic-analyzer.com/quantum-analyzer-ah-q8.html.

2. Daniel P. Thurs and Ronald L. Numbers, "Science, Pseudoscience, and Science Falsely So-Called," in *Philosophy of Pseudoscience*, ed. Massimo Pigliucci and Maarten Boudry (Chicago: University of Chicago Press, 2013), 125.

3. Kendrick Frazier, "First Contact: The News Event and the Human Response," in *Extraterrestrial Intelligence: The First Encounter*, ed. James L. Christian, 73–88 (Amherst, NY: Prometheus Books, 1976).

4. National Science Board, "Science and Technology: Public Attitudes and Understanding," in *Science and Engineering Indicators 2014* (Arlington, VA: National Science Foundation, 2014).

5. See James E. Alcock, "Extrasensory Perception," in *The Encyclopedia of the Paranormal*, ed. Gordon Stein, 241–54 (Amherst, NY: Prometheus Books, 1996); James E. Alcock, *Science and Supernature: A Critical Appraisal of Parapsychology* (Amherst, NY: Prometheus Books, 1990); Ray Hyman, *The Elusive Quarry: A Scientific Appraisal of Psychical Research* (Amherst, NY: Prometheus Books, 1989); Arthur S. Reber and James E. Alcock, "Why Parapsychological Claims Cannot Be True," *Skeptical Inquirer* 43, no. 4 (July/August 2019): 8–10, https://skepticalinquirer.org/2019/07/why-parapsychological-claims-cannot-be

-true/; Richard Wiseman, *Paranormality: Why We See What Isn't There* (London: Macmillan, 2011).

6. Ryan Shaffer and Agatha Jadwiszczok, "Psychic Defective: Sylvia Browne's History of Failure," *Skeptical Inquirer* 34, no. 2 (March/April 2010): 38–42, https://skepticalinquirer.org/2020/03/psychic-defective-sylvia-brownes-history-of-failure/; Ryan Shaffer, "The Psychic Defective Revisited: Years Later, Sylvia Browne's Accuracy Remains Dismal," *Skeptical Inquirer* 37, no. 5 (September/October 2013): 30–35 http://www.csicop.org/si/show/the_psychic_defective_revisited_years_later_sylvia_brownes_accuracy_remains/.

7. Reber and Alcock, "Why Parapsychological Claims."

Chapter 4

1. Mark Oswald, "Shirley MacLaine Selling Abiquiu Ranch," *Albuquerque Journal*, April 19, 2014, C1, https://www.abqjournal.com/386415/maclaine-selling-ranch.html.

2. Martin Gardner, *In the Name of Science* (New York: Putnam, 1952).

3. Gardner, *Name of Science*.

4. Gardner, *Name of Science*.

5. Gardner, *Name of Science*, 6–7.

6. Carl Sagan, "Wonder and Skepticism," *Skeptical Inquirer* 19, no. 1 (January/February 1995): 24–30, reprinted in *Encounters with the Paranormal: Science, Knowledge, and Belief*, ed. Kendrick Frazier (Amherst, NY: Prometheus Books, 1998).

7. Sagan, "Wonder and Skepticism."

8. Neil deGrasse Tyson, interview, *Syracuse Post-Standard*, April 19, 2009.

9. Kendrick Frazier, "Neil deGrasse Tyson on Space, Reality, Pop Culture, Skeptics, and Atheists," *Skeptical Inquirer* 45, no. 6 (September/October 2021): 6–7.

10. Piers Paul Read, *Alive: The Story of the Andes Survivors* (Philadelphia: Lippincott, 1974), 315–16.

11. Benjamin Radford, "Psychic Defective: Sylvia Browne Blunders Again," *Skeptical Inquirer* 37, no. 3 (July/August 2013): 7–8; Ryan Shaffer, "The Psychic Defective Revisited," *Skeptical Inquirer* 37, no. 5 (September/October 2013): 30–35, September/October, http://www.csicop.org/si/show/the_psychic_defective_revisited_years_later_sylvia_brownes_accuracy_remains/.

12. Arthur S. Reber and James E. Alcock, "Searching for the Impossible: Parapsychology's Elusive Quest," *American Psychologist* 75 (2020): 391–99; Arthur S. Reber and James E. Alcock, "Why Parapsychological Claims Cannot Be

True," *Skeptical Inquirer* 43, no. 4 (July/August 2019): 8–10, https://skeptical inquirer.org/2019/07/why-parapsychological-claims-cannot-be-true/.

13. Benjamin Radford, "Intuition Nearly Kills Hiker," *A Skeptic Reads the Newspaper* (blog), Center for Inquiry, July 19, 2019, https://centerforinquiry.org /blog/intuition-nearly-kills-hiker/.

14. Radford, "Intuition Nearly Kills Hiker."

15. Radford, "Intuition Nearly Kills Hiker."

16. Select Comm. on Aging, *Quackery: A $10 Billion Scandal*, H.R. Doc. 98-262, at 184 (1984).

17. Select Comm. on Aging, at 3, iii.

18. Select Comm. on Aging, at v.

19. Select Comm. on Aging, at 186.

20. Select Comm. on Aging, at v.

21. Select Comm. on Aging, at 3.

22. Select Comm. on Aging, at 4.

23. David M. Eisenberg, Ronald C. Kessler, Cindy Foster, Frances E. Norlock, David R. Calkins, and Thomas L. Delbanco, "Unconventional Medicine in the United States," *New England Journal of Medicine* 328, no. 4 (January 28, 1993): 246, http://www.nejm.org/doi/pdf/10.1056/NEJM199301283280406.

24. Eisenberg, et al., "Unconventional Medicine."

25. Tim Farley, *What's the Harm?* (blog), accessed April 7, 2023, http://whats theharm.net/index.html.

26. Bob Ladendorf and Brett Ladendorf, "Wildlife Apocalypse: How Myths and Superstitions Are Driving Animal Extinctions," *Skeptical Inquirer* 42, no. 4 (July/August 2018): 30–39.

27. Arthur W. Galston and Clifford L. Slayman, "The Not-So-Secret Life of Plants," *American Scientist* 67 (May 1979): 337–44, reprinted as "Plant Sensitivity and Sensation," in *Science and the Paranormal: The Existence of the Supernatural*, ed. George O. Abell and Barry Singer (New York: Scribner's, 1981).

28. Kenneth A. Horowitz, Donald G. Lewis, and Edgar L. Gasteiger, "Plant 'Primary Perception': Electrophysiological Unresponsiveness to Brine Shrimp Killing," *Science* 189 (1976): 478–80.

29. Daniel Chamovitz, *What a Plant Knows: A Field Guide to the Senses* (New York: Scientific American Books, 2012).

30. Merlin Sheldrake, *Entangled Life: How Fungi Make Our Worlds, Change Our Minds, and Shape Our Futures* (New York: Random House, 2020); Suzanne Simard, *Finding the Mother Tree: Discovering the Wisdom of the Forest* (New York: Alfred A. Knopf, 2021); Peter Wohlleben, *The Hidden Life of Trees: What They Feel, How They Communicate: Discoveries from a Secret World*, trans. Jane Billinghurst (Vancouver, BC: David Suzuki Institute, 2016).

31. Michael Pollan, "The Intelligent Plant," *New Yorker*, December 23 and 30, 2013, 92.

32. Ann Druyan, *Cosmos: Possible Worlds* (Washington, DC: National Geographic, 2020); *Cosmos: A Spacetime Odyssey*, directed by Brannon Braga, Bill Pope, and Ann Druyan, written by Ann Druyan and Steven Soter, aired March 9–June 8, 2014, on National Geographic Channel.

33. Richard Powers, *The Overstory* (New York: W. W. Norton, 2018), 114.

34. Powers, *Overstory*, 126.

35. Powers, *Overstory*, 142.

36. David Morrison, "Velikovsky at Fifty: Cultures in Collision on the Fringes of Science," *Skeptic* 9, no. 1 (2001).

37. Kendrick Frazier, "The Velikovsky Affair: The Distortions Continue," *Skeptical Inquirer* (Fall 1980), republished in *Paranormal Borderlands of Science*, ed. Kendrick Frazier (Amherst, NY: Prometheus Books, 1981).

38. "Velikovsky: The Controversy Continues," *New York Times Book Review*, April 20, 1980.

39. Frazier, "Velikovsky Affair."

40. Luis W. Alvarez, Walter Alvarez, Frank Asaro, and Helen V. Michel, "Extraterrestrial Cause for the Cretaceous-Tertiary Extinction," *Science* 208, no. 4448 (June 6, 1980): 1095–1108, https://doi.org/10.1126/science.208.4448.1095.

41. Morrison, "Velikovsky at Fifty."

42. Marcello Truzzi, "Pseudoscience," in *The Encyclopedia of the Paranormal*, ed. Gordon Stein, 560–75 (Amherst, NY: Prometheus Books, 1996).

43. Larry Lauden, "The Demise of the Demarcation Problem," in *But Is It Science?* ed. Michael Ruse (Amherst, NY: Prometheus Books, 1996), 349.

44. Michael D. Gordin, *The Pseudoscience Wars: Immanuel Velikovsky and the Birth of the Modern Fringe* (Chicago: University of Chicago Press, 2012).

45. For example, see Mario Bunge, "The Philosophy behind Pseudoscience," *Skeptical Inquirer* 30, no. 4 (July/August 2006): 29–37; and Mario Bunge, "What Is Pseudoscience?" *Skeptical Inquirer* 9, no. 1 (Fall 2012): 36–55.

46. Massimo Pigliucci, *Nonsense on Stilts: How to Tell Science from Bunk*, 2nd ed. (Chicago: University of Chicago Press, 2018); Massimo Pigliucci and Maarten Boudry, eds., *Philosophy of Pseudoscience: Reconsideration of the Demarcation Problem* (Chicago: University of Chicago Press, 2013).

47. Philip Kitcher, *Abusing Science: The Case against Creationism* (Cambridge, MA: MIT Press, 1984).

48. Truzzi, "Pseudoscience."

49. Martin Gardner, *Science: Good, Bad, and Bogus* (Amherst, NY: Prometheus Books, 1981).

Chapter 5

1. See *Science News*'s "Celebrating 100 Years of Science Journalism" special issue of March 26, 2022, especially the cover article by Maria Temming, "100 Years of *Science News*," *Science News* 201, no. 6 (March 26, 2022): 16–25.

2. Donald H. Menzel, *Flying Saucers* (Cambridge, MA: Harvard University Press, 1953).

3. Donald H. Menzel and Ernest H. Taves, *The UFO Enigma: The Definitive Explanation of the UFO Phenomenon* (Garden City, NY: Doubleday, 1977).

4. Philip J. Klass, *UFOs Explained* (New York: Random House, 1974), 352.

5. See Mick West, "Great Expectations: House Hearing on UAP Excites Fans, Offers Little Else," *Skeptical Inquirer* 46, no. 5 (September/October 2022).

6. Baruch Spinoza, *Tractatus Politicus* (1677), chap. 1, sect. 4, in *The Oxford Dictionary of Quotations*, 3rd ed. (Oxford, UK: Oxford University Press, 1979), 517.

7. See David Robert Grimes, *Good Thinking: Why Flawed Logic Puts Us All at Risk and How Critical Thinking Can Save the World* (New York: The Experiment, 2021), and David Robert Grimes, "Schrödinger's Bin Laden: The Irrational World of Motivated Reasoning," *Skeptical Inquirer* 46, no. 1 (January/February 2022): 34–39, for excellent recent discussions.

8. Jeffrey S. Victor, "The Social Dynamics of Conspiracy Rumors: From Satanic Panic to QAnon," *Skeptical Inquirer* 46, no. 4 (July/August 2022): 37–41.

9. James Randi, *Conjuring* (New York: St. Martin's Press, 1992), xi.

10. Harry Houdini, *Miracle Mongers and Their Methods: A Complete Exposé* (Amherst, NY: Prometheus Books, 1981).

11. Martin Gardner, "Notes of a Psi-Watcher: Lessons of a Landmark PK Hoax," *Skeptical Inquirer* 7, no. 4 (Summer 1983): 16–19, reprinted in Martin Gardner, *The New Age: Notes of a Fringe Watcher* (Buffalo, NY: Prometheus Books, 1988).

12. Joe Nickell, *Inquest on the Shroud of Turin* (Amherst, NY: Prometheus Books, 1983).

13. P. E. Damon, D. J. Donahue, B. H. Gore, A. L. Hatheway, A. J. T. Jull, T. W. Linick, P. J. Sercel, et al., "Radiocarbon Dating of the Shroud of Turin," *Nature* 337, no. 6208 (February 16, 1989): 611–15; Joe Nickell, *The Science of Miracles: Investigating the Incredible* (Amherst, NY: Prometheus Books, 2013).

14. Joseph A. Bauer, "A Surgeon's View of the 'Alien Autopsy,'" *Skeptical Inquirer* 20, no. 1 (January/February 1996); Philip J. Klass, "The MJ-12 Crashed-Saucer Documents," *Skeptical Inquirer* 12, no. 2 (Winter 1987/1988); Joe Nickell, "'Alien Autopsy' Hoax," *Skeptical Inquirer* 19, no. 6 (November/December 1995); Trey Stokes, "How to Make an 'Alien' for 'Autopsy,'" *Skeptical Inquirer* 20, no.

1 (January/February 1996); and David E. Thomas, "The 'Roswell Fragment'—Case Closed," *Skeptical Inquirer* 20, no. 6 (November/December 1996). All these articles are reprinted in Kendrick Frazier, Barry Karr, and Joe Nickell, eds., *The UFO Invasion: The Roswell Incident, Alien Abductions, and Government Coverups* (Amherst, NY: Prometheus Books, 1997). For a seventy-year update on Roswell, see Kendrick Frazier, "The Roswell Incident at 70: Facts, Not Myths," *Skeptical Inquirer* 41, no. 6 (November/December 2017).

15. J. P. Cahn, "The Flying Saucers and the Mysterious Little Men," *True*, (September 1952): 17–19, 102–12; Curtis Peebles, *Watch the Skies! A Chronicle of the Flying Saucer Myth* (Washington, DC: Smithsonian Institution Press, 1994); Benjamin Radford, *Mysterious New Mexico: Miracles, Magic, and Monsters in the Land of Enchantment* (Albuquerque: University of New Mexico Press, 2014); Robert Sheaffer, "Aztec Saucer Crash Story Rises from the Dead?" *Skeptical Inquirer* 36, no. 6 (November/December 2012); David E. Thomas, "The Aztec UFO Symposium: How This Saucer Story Started as a Con Game," *Skeptical Inquirer* 22, no. 5 (September/October 1998): 12–13.

16. Alanna Durkin Richer, "Explainer: Hundreds Charged with Crimes in Capitol Attack," Associated Press, June 7, 2022, https://apnews.com/article/capitol-siege-merrick-garland-government-and-politics-conspiracy-crime-c2e427dc0fa16077d7fb98c06e61149f.

17. For two informed reports on QAnon, see Stephanie Kemmerer, "Life, the Quniverse, and Everything, Part 1," *Skeptical Inquirer* 45, no. 2 (March/April 2021), and Stephanie Kemmerer, "Life, the Quniverse, and Everything, Part 2: QManTrafficking and the 'Plandemic,'" *Skeptical Inquirer* 45, no. 3 (May/June 2021). See also Mike Rothschild, *The Storm Is upon Us: How QAnon Became a Movement, Cult, and Conspiracy Theory of Everything* (Brooklyn, NY: Melville House, 2021).

18. Annenberg Public Policy Center, "Millions Embrace Covid-19 Misinformation, Which Is Linked to Hesitancy on Vaccination and Boosters," University of Pennsylvania, December 17, 2021, https://www.annenbergpublicpolicycenter.org/millions-embrace-covid-19-misinformation-which-is-linked-to-hesitancy-on-vaccination-and-boosters/.

19. David Byler and Yan Wu, "Opinion: Will You Fall into the Conspiracy Theory Rabbit Hole? Take Our Quiz and Find Out," *Washington Post*, October 6, 2021, https://www.washingtonpost.com/opinions/interactive/2021/conspiracy-theory-quiz/.

20. Massimo Pigliucci, "Science Denialism Is a Form of Pseudoscience," *Skeptical Inquirer* 46, no. 1 (January/February 2022): 19–20.

Chapter 6

1. Harriet Hall, *Women Aren't Supposed to Fly: The Memoirs of a Female Flight Surgeon* (N.p.: iUniverse, 2008).

2. Harriet Hall, "Science Envy in Alternative Medicine," *Skeptical Inquirer* 43, no. 4 (July/August 2019): 21–23.

3. Hall, "Science Envy."

4. Hall, "Science Envy."

5. Paul A. Offit, *Do You Believe in Magic? The Sense and Nonsense of Alternative Medicine* (New York: HarperCollins, 2013).

6. Edzard Ernst, *So-Called Alternative Medicine (SCAM) for Cancer* (Cham, Switzerland: Springer, 2021), 4–5.

7. The other is Edzard Ernst, *SCAM: So-Called Alternative Medicine* (London: Imprint Academic, 2018).

8. Ernst, *SCAM for Cancer*, 5.

9. Ernst, *SCAM for Cancer*, 6.

10. Ernst, *SCAM for Cancer*, 6–7.

11. Ernst, *SCAM Cancer*, 7.

12. For more on her quest, see the short documentary *Lady Ganga: Nilza's Story* (Lumiere Media, 2015).

13. Ernst, *SCAM: So-Called Alternative Medicine.*

14. Ernst, *SCAM: So-Called Alternative Medicine.*

15. Ernst, *SCAM: So-Called Alternative Medicine.*

16. Edzard Ernst, Max H. Pittler, Clare Stevinson, and Adrian White, eds., *The Desktop Guide to Complementary and Alternative Medicine* (Edinburgh, UK: Mosby, 2001).

17. Ernst, *SCAM: So-Called Alternative Medicine.*

18. Ernst, *SCAM: So-Called Alternative Medicine.*

19. Quoted in Edzard Ernst, *Charles, the Alternative Prince: An Unauthorised Biography* (Exeter, UK: Imprint Academic, 2022), and Harriet Hall, "The Prince of Alternative Medicine: Review of Edzard Ernst, *Charles, the Alternative Prince*," *Skeptical Inquirer* 46, no. 3 (May/June 2022): 59–61.

20. Hall, "Prince of Alternative Medicine."

21. Ernst, *SCAM: So-Called Alternative Medicine.*

22. See Edzard Ernst, *Edzard Ernst* (blog), accessed April 11, 2023, https://edzardernst.com.

23. Sense about Science, "Maddox Prize 2015," accessed April 11, 2023, https://senseaboutscience.org/activities/2015-john-maddox-prize/.

24. Ernst, *SCAM: So-Called Alternative Medicine.*

25. Ernst, *SCAM: So-Called Alternative Medicine.*

26. Ernst, *SCAM for Cancer*, 8–13.
27. Ernst, *SCAM for Cancer*, 10.
28. Ernst, *SCAM Cancer*, 12.
29. Ernst, *SCAM for Cancer*, 12.
30. Ernst, *SCAM Cancer*, 12.
31. Ernst, *Charles, the Alternative Prince*.
32. Hall, "Prince of Alternative Medicine."
33. Hall, "Prince of Alternative Medicine."
34. Paul Offit, *Bad Advice: Or Why Celebrities, Politicians, and Activists Aren't Your Best Source of Health Information* (New York: Columbia University Press, 2018).
35. Offit, *Bad Advice*.
36. Offit, *Bad Advice*.
37. Offit, *Bad Advice*.
38. For more, see my extended comment about Oprah quoted in Martin Gardner, "Oprah Winfrey: Bright (but Gullible) Billionaire," *Skeptical Inquirer* 34, no. 2 (March/April 2010): 55–56.
39. Offit, *Do You Believe in Magic?*
40. Offit, *Do You Believe in Magic?*
41. Offit, *Do You Believe in Magic?*
42. Gardner, "Oprah Winfrey."
43. Gardner, "Oprah Winfrey."
44. Gardner, "Oprah Winfrey."
45. Gardner, "Oprah Winfrey."
46. Bernardo Montes de Oca, "What Is Goop? Here's Why Everyone Hates Gwyneth Paltrow's Company," Slidebean, September 2, 2021, https://slidebean.com/story/what-is-goop-paltrow-slidebean#:~:text=Goop%20claimed%20that%20its%20Body,said%20it%20was%20an%20error.
47. Goop, "What's Goop?" accessed April 11, 2023, https://goop.com/whats-goop/.
48. Julie Mazziotta, "Goop to Pay $145,000 Settlement for Misleading Claims about the Effectiveness of Vaginal Eggs," *People*, September 5, 2018, https://people.com/health/gwyneth-paltrow-goop-pay-settlement-misleading-claims-vaginal-eggs/.
49. Montes de Oca, "What Is Goop?"
50. Quoted in Alene Tchekmedyian, "Gwyeth Paltrow's Goop to Offer Refunds over 'Unsubstantiated' Claims about Health Benefits," *Los Angeles Times*, September 4, 2018, https://www.latimes.com/local/lanow/la-me-ln-goop-settlement-20180904-story.html.
51. Montes de Oca, "What Is Goop?"

52. Montes de Oca, "What Is Goop?"
53. Robyn Blumner, "Why Full of Bull? Here's Why: Postcards from Reality Column," *Skeptical Inquirer* 45, no. 6 (November/December 2021): 14–15.
54. Harriet Hall, "The Care and Feeding of the Vagina," *Skeptical Inquirer* 42, no. 5 (September/October 2018): 28–29.
55. Hall, "Care and Feeding."
56. Tracy Smith, "In Conversation: Gwyneth Paltrow," *CBS News Sunday Morning*, aired September 25, 2022.
57. Smith, "In Conversation."
58. Timothy Caulfield, *Is Gwyneth Paltrow Wrong about Everything? How the Famous Sell Us Elixirs of Health, Beauty and Happiness* (Boston: Beacon Press, 2015), xv.
59. Caulfield, *Is Gwyneth Paltrow Wrong*, 5.
60. Caulfield, *Is Gwyneth Paltrow Wrong*, 7.
61. Caulfield, *Is Gwyneth Paltrow Wrong*.
62. Caulfield, *Is Gwyneth Paltrow Wrong*, 9–10.
63. Caulfield, *Is Gwyneth Paltrow Wrong*, xi.
64. Offit, *Do You Believe in Magic?*

Chapter 7

1. Richard Feynman, *What Do* You *Care What Other People Think?* (New York: Norton, 1988).
2. Isaac Asimov, *Past, Present, and Future* (Buffalo, NY: Prometheus Books, 1987), 65.
3. Kendrick Frazier, *Our Turbulent Sun: The Scientific Quest for a New Understanding of Our Sun and Its Surprising Irregularities, Variations, and Effects* (Englewood Cliffs, NJ: Prentice-Hall, 1982).
4. Richard Dawkins, "Science, Delusion, and the Appetite for Wonder," Edge, 1996, http://edge.org/conversation/science-delusion-and-the-appetite-for-wonder.
5. Rebecca Goldstein, keynote address ("Reason for Change" conference, Center for Inquiry, Amherst, New York, June 11–15, 2015).
6. Eugenie Scott, presentation ("Reason for Change" conference, Center for Inquiry, Amherst, New York, June 11–15, 2015).
7. J. Bronowski, *The Common Sense of Science* (Cambridge, MA: Harvard University Press, 1976).
8. Gerard M. Verschuuren, *Life Scientists: Their Convictions, Their Activities, Their Values* (North Andover, MA: Genesis, 1995).

9. Verschuuren, *Life Scientists.*

10. Verschuuren, *Life Scientists.*

11. Event Horizon Telescope Collaboration, et al., "First M87 Event Horizon Telescope Results. I. The Shadow of the Supermassive Black Hole," *Astrophysical Journal Letters* 875, no. 1 (April 10, 2019), https://iopscience.iop.org/article/10.3847/2041-8213/ab0ec7.

12. Martin J. P. Sullivan, Simon L. Lewis, Kofi Affum-Baffoe, Carolina Castilho, Flávia Costa, Aida Cuni Sanchez, Corneille E. N. Ewango, et al., "Long-Term Thermal Sensitivity of Earth's Tropical Forests," *Science* 368, no. 6493 (May 22, 2020): 869–74.

13. Maya Ajmera, "Conversations with Maya," interview by Thomas Rosenbaum, *Science News* 195, no. 6 (March 30, 2019): 28–29.

14. Richard Berthold, lecture (meeting, New Mexicans for Science and Reason, April 11, 2007).

15. Berthold, lecture. See also Richard M. Berthold, *Dare to Struggle: The History and Society of Ancient Greece* (N.p.: iUniverse, 2009).

A number of recent books explore science's origins in ancient Greek times. Among them are Stephen Bertman, *The Genesis of Science: The Story of Greek Imagination* (Amherst, NY: Prometheus Books, 2010); Demetris Nicolaides, *In the Light of Science: Our Ancient Quest for Knowledge and the Measure of Modern Physics* (Amherst, NY: Prometheus Books, 2014); Richard H. Schlagel, *Seeking the Truth: How Science Has Prevailed over the Supernatural Worldview* (Amherst, NY: Humanity Books, 2011); and Nobel laureate physicist Steven Weinberg, *To Explain the World: The Discovery of Modern Science* (New York: HarperCollins, 2015), which also emphasizes the shortcomings in Weinberg's insights.

16. Quoted in Will Durant and Ariel Durant, *The Age of Reason Begins*, pt. 7, *The Story of Civilization* (New York: Simon and Schuster, 1961), 175.

17. Durant and Durant, *Age of Reason Begins*, 182.

18. Durant and Durant, *Age of Reason Begins.*

19. Peter Medawar, "Two Conceptions of Science," in *Pluto's Republic* (Oxford, UK: Oxford University Press, 1984), 31.

20. Medawar, "Two Conceptions of Science," 30.

21. Medawar, "Two Conceptions of Science," 32

22. Medawar, "Two Conceptions of Science," 33.

23. Albert Einstein, "Imagination is more important than knowledge," Quote Investigator, January 1, 2013, http://quoteinvestigator.com/2013/01/01/einstein-imagination/.

24. I have a copy of this letter of Einstein's, sent to me by its recipient, a reader of *Science News*, in the 1970s.

25. Richard P. Feynman, "Solutions of Maxwell's Equations in Free Space," in *The Feynman Lectures on Physics*, by Richard P. Feynman, Robert B. Leighton, and Matthew Sands, vol. 2, *Mainly Electromagnetism and Matter* (Redwood City, CA: Addison-Wesley, 1989), https://www.feynmanlectures.caltech.edu/.

26. Feynman, "Solutions of Maxwell's Equations."

27. Medawar, "Two Conceptions of Science."

28. Max Planck, *Scientific Biography and Other Papers: With a Memorial Address on Max Planck* (New York: Philosophical Library, 1949).

29. UC Museum of Paleontology, "Prepare and Plan: Correcting Misconceptions," University of California, Berkeley, 2023, http://undsci.berkeley.edu/teaching/misconceptions.php#b1.

30. UC Museum of Paleontology, "Prepare and Plan."

31. This is the definition used in by the National Academy of Sciences. See National Academy of Sciences Institute of Medicine, *Science, Evolution, and Creationism* (Washington, DC: National Academies Press, 2008), 10.

32. National Academy of Sciences Institute of Medicine, *Science, Evolution, and Creationism*, 11.

Chapter 8

1. Charles M. Wynn and Arthur W. Wiggins, *Quantum Leaps in the Wrong Direction: Where Real Science Ends—and Pseudoscience Begins*, 2nd ed., with cartoons by Sydney Harris (New York: Oxford University Press, 2017).

2. Mario Bunge, *Between Two Worlds: Memoirs of a Philosopher-Scientist* (New York: Springer Berlin Heidelberg, 2016); Mario Bunge, *From a Scientific Point of View: Reasoning and Evidence Beat Improvisation across Fields* (Newcastle upon Tyne, UK: Cambridge Scholars, 2018); Michael R. Matthews, ed., *Mario Bunge: A Centenary Festschrift* (Cham, Switzerland: Springer, 2019).

3. Michael R. Matthews, "Mario Bunge: Physicist, Philosopher, Champion of Science, Citizen of the World," *Skeptical Inquirer* 44, no. 4 (July/August 2020): 7–8.

4. Mario Bunge, "The Philosophy behind Pseudoscience," *Skeptical Inquirer* 30, no. 4 (July/August 2006): 29–37; Mario Bunge, "What Is Pseudoscience?" *Skeptical Inquirer* 9, no. 1 (Fall 1984): 36–46.

5. Bunge, "What Is Pseudoscience?" 41.

6. Bunge, "What Is Pseudoscience?" 41–42.

7. Bunge, "What Is Pseudoscience?" 44. One philosopher, Maarten Boudry, to whom I showed this paragraph, says he thinks it's a bit too simple. In an email to me from him on September 21, 2017, he said, "Scientists tend to be quite

conservative and skeptical of radical ideas, often for reasons. . . . But yes, in the long run science is arranged in such a way that the new ideas will prevail, if there's evidence to back them up."

8. Bunge, "What Is Pseudoscience?" 44.
9. Bunge, "What Is Pseudoscience?" 44.
10. Bunge, "What Is Pseudoscience?" 46.
11. Marcello Truzzi, "Pseudoscience," in *The Encyclopedia of the Paranormal*, ed. Gordon Stein (Amherst, NY: Prometheus Books, 1996).
12. Truzzi, "Pseudoscience."
13. Truzzi, "Pseudoscience."
14. Mario Bunge, email to the author.
15. Dana Richards, ed., *Dear Martin, Dear Marcello*: *Gardner and Truzzi on Skepticism* (Hackensack, NJ: World Scientific, 2017).
16. See the section on the Velikovsky affair in chapter 4, "What's the Harm? Why Does It Matter?"
17. Bunge, "Philosophy behind Pseudoscience."
18. Kendrick Frazier, "Editorial: Explanatory Frameworks and Investigative Exposés," *Skeptical Inquirer* 30, no. 4 (July/August 2006): 4.
19. Bunge, "Philosophy behind Pseudoscience."
20. Bunge, "Philosophy behind Pseudoscience."
21. Bunge, "Philosophy behind Pseudoscience."
22. Bunge, "Philosophy behind Pseudoscience."
23. Bunge, "Philosophy behind Pseudoscience."
24. Bunge, "Philosophy behind Pseudoscience."
25. Bunge, "Philosophy behind Pseudoscience."
26. Massimo Pigliucci and Maarten Boudry, eds., *Philosophy of Pseudoscience: Reconsideration of the Demarcation Problem* (Chicago: University of Chicago Press, 2013).
27. Massimo Pigliucci, "Science Denialism Is a Form of Pseudoscience," *Skeptical Inquirer* 46, no. 1 (January/February 2022): 19–20. If you want a further introduction to him, I suggest reading several of his *Skeptical Inquirer* columns or his excellent book *Nonsense on Stilts: How to Tell Science from Bunk*, 2nd ed. (Chicago: University of Chicago Press, 2018), or checking him out online.
28. Larry Laudan, "The Demise of the Demarcation Problem," in *Physics, Philosophy and Psychoanalysis: Essays in Honor of Adolf Grünbaum*, ed. R. S. Cohen and L. Lauden (Dordrecht, Holland: D. Reidel, 1983), 111–27.
29. Massimo Pigliucci and Maarten Boudry, introduction to *Philosophy of Pseudoscience: Reconsideration of the Demarcation Problem*, ed. Massimo Pigliucci and Maarten Boudry (Chicago: University of Chicago Press, 2013).
30. Pigliucci and Boudry, introduction.

31. Pigliucci and Boudry, introduction.
32. Pigliucci and Boudry, introduction.
33. Massimo Pigliucci, "The Demarcation Problem: A (Belated) Response to Laudan," in *Philosophy of Pseudoscience: Reconsideration of the Demarcation Problem*, ed. Massimo Pigliucci and Maarten Boudry (Chicago: University of Chicago Press, 2013).
34. Pigliucci, "Demarcation Problem."
35. Pigliucci, "Demarcation Problem."
36. Pigliucci, "Demarcation Problem."
37. Pigliucci, "Demarcation Problem."
38. Martin Mahner, "Science and Pseudoscience," in *Philosophy of Pseudoscience: Reconsideration of the Demarcation Problem*, ed. Massimo Pigliucci and Maarten Boudry (Chicago: University of Chicago Press, 2013).
39. Mahner, "Science and Pseudoscience."
40. Mahner, "Science and Pseudoscience."
41. Mahner, "Science and Pseudoscience."
42. Sven Ove Hansson, "Defining Pseudoscience and Science," in *Philosophy of Pseudoscience: Reconsideration of the Demarcation Problem*, ed. Massimo Pigliucci and Maarten Boudry (Chicago: University of Chicago Press, 2013).
43. Mahner, "Science and Pseudoscience."
44. Hansson, "Defining Pseudoscience and Science."
45. James Ladyman, "Toward a Demarcation of Science from Pseudoscience," in *Philosophy of Pseudoscience: Reconsideration of the Demarcation Problem*, ed. Massimo Pigliucci and Maarten Boudry (Chicago: University of Chicago Press, 2013).
46. Harry G. Frankfurt, *On Bullshit* (Princeton, NJ: Princeton University Press, 2005), quoted in Ladyman, "Toward a Demarcation."
47. Carl T. Bergstrom and Jevin D. West, *Calling Bullshit: The Art of Skepticism in a Data-Driven World* (New York: Random House, 2020); Stephen Law, *Believing Bullshit: How Not to Get Sucked into an Intellectual Black Hole* (Amherst, NY: Prometheus Books, 2011).
48. Ladyman, "Toward a Demarcation."
49. Noretta Koertge, "Belief Buddies versus Critical Communities," in *Philosophy of Pseudoscience: Reconsideration of the Demarcation Problem*, ed. Massimo Pigliucci and Maarten Boudry (Chicago: University of Chicago Press, 2013).
50. Koertge, "Belief Buddies"; emphasis mine.
51. Koertge, "Belief Buddies."
52. Koertge, "Belief Buddies."
53. Koertge, "Belief Buddies."
54. Koertge, "Belief Buddies."

55. Susan Haack has written a number of insightful books, like *Defending Science—within Reason: Between Scientism and Cynicism* (Amherst, NY: Prometheus Books, 2003), where her discussions about the scientific method can be found; *Manifesto of a Passionate Moderate: Unfashionable Essays* (Chicago: University of Chicago Press, 1998); and *Philosophy of Logics* (Cambridge, UK: Cambridge University Press, 1978).

56. Susan Haack, *Putting Philosophy to Work: Inquiry and Its Place in Culture: Essays on Science, Religion, Law, Literature, and Life* (Amherst, NY: Prometheus Books, 2013).

Chapter 9

1. For a very good analysis of why and how that happened, see Jeffrey S. Victor, "The Social Forces behind Trump's War on Science," *Free Inquiry* 38, no. 5 (August/September 2018).

2. Kendrick Frazier, *The Violent Face of Nature: Severe Phenomena and Natural Disasters* (New York: Morrow, 1979).

3. I've adapted this definition from British climatologist Hubert H. Lamb, as stated in Walter Orr Roberts and Henry Lansford, *The Climate Mandate* (San Francisco: W. H. Freeman, 1979), 18.

4. Roger Revelle and Hans E. Suess, "Carbon Dioxide Exchange between Atmosphere and Ocean and the Question of an Increase of Atmospheric CO_2 during the Past Decades," *Tellus* 9, no. 1 (February 1957): 19.

5. Roger Revelle, "Carbon Dioxide and World Climate," *Scientific American* 247, no. 2 (August 1982).

6. Earth Observatory, "Roger Revelle (1909–1991)," June 19, 2000, https://earthobservatory.nasa.gov/Features/Revelle/revelle_3.php.

7. Stephen H. Schneider, *Global Warming: Are We Entering the Greenhouse Century?* (San Francisco: Sierra Club Books, 1989); Stephen H. Schneider and Randi Londer, *The Coevolution of Climate and Life* (San Francisco: Sierra Club Books, 1984); Stephen H. Schneider, with Lynne E. Mesirow, *The Genesis Strategy: Climate and Global Survival* (New York: Plenum Press, 1976).

8. Stephen H. Schneider, *Science as a Contact Sport: Inside the Battle to Save Earth's Climate* (Washington, DC: National Geographic, 2009).

9. Schneider, *Science as a Contact Sport.*

10. See Robert Clairborne, *Climate, Man, and History* (New York: Norton, 1970); Brian Fagan, *The Long Summer: How Climate Changed Civilization* (New York: Basic Books, 2004). You might also want to read Brian Fagan, *The Little Ice Age: How Climate Made History, 1300–1850* (New York: Basic Books, 2000).

11. John Imbrie and Katherine Palmer Imbrie, *The Ice Ages: Solving the Mystery* (Short Hills, NJ: Enslow, 1979); Kendrick Frazier, "The Dance of the Orbits," in *Our Turbulent Sun: The Scientific Quest for a New Understanding of Our Sun and Its Surprising Irregularities, Variations, and Effects* (Englewood Cliffs, NJ: Prentice-Hall, 1982).

12. J. D. Hays, John Imbrie, and N. J. Shackleton, "Variations in the Earth's Orbit: Pacemaker of the Ice Ages," *Science* 194, no. 4270 (December 10, 1976): 1121–32.

13. Henry Pollack, *A World without Ice* (New York: Avery/Penguin Group, 2009).

14. Robert Henson, *The Rough Guide to Climate Change* (London: Rough Guides, 2008).

15. Henry Pollack, lecture (meeting, New Mexicans for Science and Reason December 2, 2009). See also Henson, *Rough Guide.*

16. Small-M, "Math! How Much CO_2 by Weight in the Atmosphere?" March 30, 2007, https://micpohling.wordpress.com/2007/03/30/math-how-much-co2-by-weight-in-the-atmosphere/.

17. National Oceanic and Atmospheric Administration, "Carbon Dioxide Now More Than 50% Higher Than Pre-industrial Levels," June 3, 2022, https://www.noaa.gov/news-release/carbon-dioxide-now-more-than-50-higher-than-pre-industrial-levels.

18. National Oceanic and Atmospheric Administration, "Carbon Dioxide Now More."

19. *Cosmos: A Spacetime Odyssey*, directed by Brannon Braga, Bill Pope, and Ann Druyan, written by Ann Druyan and Steven Soter, aired March 9–June 8, 2014, on National Geographic Channel.

20. You can track these numbers yourself at CO2.Earth, "Earth's CO_2 Home Page," accessed April 15, 2023, https://www.co2.earth.

21. A good graphic and more information are available at National Aeronautics and Space Administration, "Graphic: The Relentless Rise of Carbon Dioxide," accessed April 15, 2023, https://climate.nasa.gov/climate_resources/24/graphic-the-relentless-rise-of-carbon-dioxide/.

22. Rebecca Lindsey, "Climate Change: Atmospheric Carbon Dioxide," National Oceanic and Atmospheric Administration, June 23, 2022, https://www.climate.gov/news-features/understanding-climate/climate-change-atmospheric-carbon-dioxide.

23. Pollack, lecture.

24. Pollack, lecture.

25. Valérie Masson-Delmotte, Panmao Zhai, Hans-Otto Pörtner, Debra Roberts, Jim Skea, Priyadarshi R. Shukla, Anna Pirani, Wilfran Moufouma-

Okia, Clotilde Péan, Roz Pidcock, et al., *Global Warming of 1.5°C: An IPCC Special Report on the Impacts of Global Warming of 1.5°C above Pre-industrial Levels and Related Global Greenhouse Gas Emission Pathways, in the Context of Strengthening the Global Response to the Threat of Climate Change, Sustainable Development, and Efforts to Eradicate Poverty* (Cambridge, UK: Cambridge University Press, 2019), 4.

26. Masson-Delmotte, et al., *Global Warming of 1.5°C*, 60.

27. Quoted in Amanda Schmidt, "UN Climate Change Panel Says 'Unprecedented Changes' Needed to Prevent Rapid Global Warming," Yahoo! News, October 8, 2018, https://www.yahoo.com/news/un-climate-change-panel-says-173744678.html.

28. Masson-Delmotte, et al., *Global Warming of 1.5°C*, 53.

29. Masson-Delmotte, et al., *Global Warming of 1.5°C*, 319.

30. Hans-Otto Pörtner, Debra C. Roberts, Melinda M. B. Tignor, Elvira Poloczanska, Katja Mintenbeck, Andrés Alegría, Marlies Craig, et al., eds., *Climate Change 2022: Impacts, Adaptation and Vulnerability: Working Group II Contribution to the Sixth Assessment Report of the Intergovernmental Panel on Climate Change* (Cambridge, UK: Cambridge University Press, 2022), 9.

31. Hans-Otto Pörtner, Debra C. Roberts, Melinda M. B. Tignor, Elvira Poloczanska, Katja Mintenbeck, Andrés Alegría, Marlies Craig, et al., eds., *Climate Change 2022: Impacts, Adaptation and Vulnerability: Working Group II Contribution to the Sixth Assessment Report of the Intergovernmental Panel on Climate Change* (Cambridge, UK: Cambridge University Press, 2022), 9.

32. Michael Le Page, "Heatwave in China Is the Most Severe Ever Recorded in the World," *New Scientist*, August 23, 2022, https://www.newscientist.com/article/2334921-heatwave-in-china-is-the-most-severe-ever-recorded-in-the-world/.

33. Quoted in Emma Newburger, "UN Warns World Is Entering 'Uncharted Territories of Destruction' from Climate Crisis," CNBC, September 13, 2022, https://www.cnbc.com/2022/09/13/world-entering-uncharted-territories-of-destruction-climate-crisis-un.html#:~:text=The%20United%20Nations%20is%20warning,to%20reduce%20greenhouse%20gas%20emissions.

34. Steve Newman, "Earthweek: Diary of a Changing World: Week Ending August 12, 2022," https://www.earthweek.com/_files/ugd/532fc1_2ccad38118d442fd8e5900ce92385aea.pdf, and Steve Newman, "Earthweek: Diary of a Changing World: Week Ending September 16, 2022," https://www.earthweek.com/_files/ugd/532fc1_30e9825882994c87aee3feeef999c3b0.pdf.

35. Alex DeMarban, "Storms Pummel Bering Sea Islands after 'Crazy' Ice Melt-Off," *Anchorage Daily News*, February 28, 2018, https://www.adn.com/alaska-news/rural-alaska/2018/02/27/warm-storms-pummel-bering-sea-lead

ing-to-crazy-ice-melt-off/#:~:text=A%20swath%20of%20sea%20ice,17%20years%20ago%2C%20scientists%20say.

36. Al Gore, *An Inconvenient Truth: The Crisis of Global Warming* (New York: Viking, 2007); *An Inconvenient Truth*, directed by Davis Guggenheim, written by Al Gore (Beverly Hills, CA: Lawrence Bender Productions, 2006).

37. Michael E. Mann and Lee Toles, *The Madhouse Effect: How Climate Change Denial Is Threatening Our Planet, Destroying Our Politics, and Driving Us Crazy* (New York: Columbia University Press, 2016).

38. Michael E. Mann, correspondence to the author, August 31, 2022.

39. Naomi Oreskes and Erik M. Conway, *Merchants of Doubt: How a Handful of Scientists Obscured the Truth on Issues from Tobacco Smoke to Global Warming* (New York: Bloomsbury Press, 2010).

40. Oreskes and Conway, *Merchants of Doubt*.

41. Oreskes and Conway, *Merchants of Doubt*.

42. Kendrick Frazier, "The Winter of Our Discontent (From the Editor)," *Skeptical Inquirer* 34, no. 3 (May/June 2010): 4; Mann and Toles, *Madhouse Effect*.

43. Kendrick Frazier, "What Exxon Knew about Arctic Warming Twenty-Five Years Ago," *Skeptical Inquirer* 40, no. 1 (January/February 2016): 5–6.

44. Oreskes and Conway, *Merchants of Doubt*.

45. Paul Thacker, "Fred Singer Has Passed. May He Rest," *Drilled News*, April 16, 2020, www.drillednews.com/post/fred-singer-obituary-climate-denier.

46. Thacker, "Fred Singer Has Passed."

Chapter 10

1. Immanuel Velikovsky, *Worlds in Collision* (New York: Macmillan, 1950). For more about Velikovsky, see chapter 4.

2. Philip J. Klass, *UFOs Explained* (New York: Random House, 1974).

3. Lawrence David Kusche, *The Bermuda Triangle Mystery—Solved* (New York: Harper and Row, 1975).

4. The Amazing [James] Randi, *The Magic of Uri Geller* (New York: Ballantine Books, 1975).

5. Bart J. Bok, Lawrence E. Jerome, Paul Kurtz, et al., "Objections to Astrology: A Statement by 186 Leading Scientists," *Humanist* 35, no. 5 (September/October 1975): 4–6.

6. Bart J. Bok, "A Critical Look at Astrology," *Humanist* 35, no. 5 (September/October 1975): 6–9.

7. Bart J. Bok and Lawrence E. Jerome, *Objections to Astrology: 192 Leading Scientists, Including 19 Nobel Prize Winners, Disavow Astrology* (Amherst, NY: Prometheus Books, 1975).

8. Martin Gardner, *Fads and Fallacies in the Name of Science* (New York: Dover, 1957).

9. Dana Richards, ed., *Dear Martin, Dear Marcello: Gardner and Truzzi on Skepticism* (Hackensack, NJ: World Scientific, 2017).

10. Richards, *Dear Martin, Dear Marcello.*

11. Richards, *Dear Martin, Dear Marcello.*

12. Paul Kurtz, ed., *Skeptical Odysseys: Personal Accounts by the World's Leading Paranormal Inquirers* (Amherst, NY: Prometheus Books, 2001).

13. Kurtz, *Skeptical Odysseys.*

14. Kurtz, *Skeptical Odysseys.*

15. Kurtz, *Skeptical Odysseys.*

16. Quoted in Kendrick Frazier, "Science and the Parascience Cults," *Science News* 109 (May 29, 1977): 346–48, 350; Kendrick Frazier, "Committee for the Scientific Investigation of Claims of the Paranormal (CSICOP)," in *The Encyclopedia of the Paranormal*, ed. Gordon Stein (Amherst, NY: Prometheus Books, 1996), 168–81.

17. Quoted in Frazier, "Science and the Parascience Cults"; Kendrick Frazier, "CSICOP."

18. Quoted in Frazier, "Science and the Parascience Cults"; Kendrick Frazier, "CSICOP."

19. Quoted in Frazier, "CSICOP."

20. Kendrick Frazier, "Challenging Pseudoscience," *Science News* 109, no. 22 (May 29, 1976).

21. Kurtz, *Skeptical Odysseys.*

22. Kurtz, *Skeptical Odysseys.*

23. Kurtz, *Skeptical Odysseys.*

24. Richards, *Dear Martin, Dear Marcello.*

25. Kurtz, *Skeptical Odysseys.*

26. Robert Blaskiewicz and Mike Jursulic, "Arthur C. Cramp: The Quackbuster Who Professionalized American Medicine," *Skeptical Inquirer* 42, no. 6 (November/December 2018): 45–50; Harry Houdini, *Miracle Mongers and Their Methods: A Complete Exposé* (Amherst, NY: Prometheus Books, 1981).

27. Daniel Loxton, "Why Is There a Skeptical Movement? Part One: Two Millennia of Paranormal Skepticism," 2013, https://www.arvindguptatoys.com/arvindgupta/skeptical-movement.pdf.

28. Loren Pankratz, *Mysteries and Secrets Revealed: From Oracles at Delphi to Spiritualism in America* (Lanham, MD: Prometheus Books, 2021).

29. For more on each, see Frazier, "CSICOP."

30. Quoted in Ron Amundson, "Follow-Up: Watson and the 'Hundredth Monkey Phenomenon,'" *Skeptical Inquirer* 11, no. 3 (Spring 1987): 303–4. See also Ron Amundson, "The Hundredth Monkey Phenomenon," *Skeptical Inquirer* 9, no. 4 (Summer 1985): 348–56.

31. Kendrick Frazier, ed., *The Hundredth Monkey and Other Paradigms of the Paranormal: A* Skeptical Inquirer *Collection* (Buffalo, NY: Prometheus Books, 1991).

32. Kendrick Frazier, ed., *Paranormal Borderlands of Science* (Buffalo, NY: Prometheus Books, 1981).

33. Kendrick Frazier, introduction to *Paranormal Borderlands of Science*, ed. Kendrick Frazier (Buffalo, NY: Prometheus Books, 1981).

34. Kendrick Frazier, ed., *Encounters with the Paranormal: Science, Knowledge, and Belief* (Amherst, NY: Prometheus Books, 1998); Frazier, *Hundredth Monkey*; Kendrick Frazier, ed., *Science Confronts the Paranormal* (Buffalo, NY: Prometheus Books, 1986); Kendrick Frazier, ed., *Science under Siege: Defending Science, Exposing Pseudoscience* (Amherst, NY: Prometheus Books, 2009).

35. Kendrick Frazier, Barry Karr, and Joe Nickell, eds., *The UFO Invasion: The Roswell Incident, Alien Abductions, and Government Coverups* (Amherst, NY: Prometheus Books, 1997).

Chapter 11

1. Paul Kurtz, ed., *Skeptical Odysseys: Personal Accounts by the World's Leading Paranormal Inquirers* (Amherst, NY: Prometheus Books, 2001).

2. Paul Kurtz, "Science and the Public: Summing Up Thirty Years of the *Skeptical Inquirer*," *Skeptical Inquirer* 30, no. 5 (September/October 2006): 13–19.

3. Kendrick Frazier, "Committee for the Scientific Investigation of Claims of the Paranormal (CSICOP)," in *The Encyclopedia of the Paranormal*, ed. Gordon Stein (Amherst, NY: Prometheus Books, 1996), 168–81, https://skepticalinquirer.org/history-of-csicop/.

4. Kendrick Frazier, "CSICOP Timeline: A Capsule History in 85 Easy Steps," *Skeptical Inquirer* 25, no. 3 (May/June 2001), http://www.skepticalinquirer.org/si/show/csicop_timeline.

5. Kendrick Frazier, ed., "Issues in Science and Skepticism," special issue, *Skeptical Inquirer* 40, no. 5 (September/October 2016); Kendrick Frazier, ed., "Odysseys in Scientific Skepticism," special issue, *Skeptical Inquirer* 40, no. 6 (November/December 2016). The timeline is on pp. 51–55 of the latter.

6. Kendrick Frazier, "It's CSI Now, Not CSICOP," *Skeptical Inquirer* 31, no. 1 (January/February 2007).

7. Frazier, "It's CSI Now."

8. Paul Kurtz, "Editorial: New Directions for Skeptical Inquiry," *Skeptical Inquirer* 31, no. 1 (January/February 2007): 7.

9. Frazier, "It's CSI Now."

10. Wendy M. Grossman and Christopher C. French, eds., *Why Statues Weep: The Best of* The Skeptic (London: Philosophy Press, 2010).

11. Grossman and French, *Why Statues Weep.*

12. Grossman and French, *Why Statues Weep.*

13. Grossman and French, *Why Statues Weep*, xi.

14. Grossman and French, *Why Statues Weep.*

15. *The Skeptic*, "About 'The Skeptic,'" accessed April 16, 2023, https://www.skeptic.org.uk/about/. See the UK Web Archive at https://www.webarchive.org.uk.

16. Kurtz, *Skeptical Odysseys*, 17.

17. Piero Angela, "The World of Galileo Galilei" (presentation, Fifth World Skeptics Congress, Abano Terme, Italy, October 8–10, 2004).

18. Barry Karr, "Skeptical Inquirer Presents Series a Big Success: Fills CSI-Con Gap," *Skeptical Inquirer* 45, no. 3, (May/June 2021).

19. "Center for Inquiry, Richard Dawkins Foundation Merging; Robyn Blumner CFI's New CEO," *Skeptical Inquirer* 40, no. 3 (May/June 2016): 5–6.

20. Barry Karr, quoted in "Magicians, Skeptics Share Their Memories of James Randi," *Skeptical Inquirer* 45, no. 1 (January/February 2021): 44.

21. Barry Karr, fund-drive letter, October 20, 2020.

22. "Committee for Skeptical Inquiry Names Ten New Fellows," *Skeptical Inquirer* 45, no. 1 (January/February 2021).

23. Paul Kurtz, "From the Chairman," *Skeptical Inquirer* 5, no. 3 (Spring 1981): 2–4.

24. Kendrick Frazier, "In Troubled Times, This Is What We Do," *Skeptical Inquirer* 42, no. 2 (March/April 2018): 14–15.

25. H. Holden Thorp, "Letters to the Editor: The State of Our Nation," *Skeptical Inquirer* 45, no. 1 (January/February 2021): 64.

Chapter 12

1. Tommaso Treu, quoted in Daniel Clery, "Webb Telescope Reveals Unpredicted Bounty of Bright Galaxies in Early Universe," *Science*, August 9, 2022,

https://www.science.org/content/article/webb-telescope-reveals-unpredicted-bounty-bright-galaxies-early-universe.

2. Ewan Callaway, "'It Will Change Everything': DeepMind's AI Makes Gigantic Leap in Solving Protein Structures," *Nature* 588, no. 7837 (December 10, 2020): 203–4; Robert F. Service, "Protein Structures for All," *Science* 374, no. 6574 (December 17, 2021): 1426–28.

3. Service, "Protein Structures for All."

4. Matthew Sparks, "DeepMind's Protein-Folding AI Cracks Biology's Biggest Problem," *New Scientist*, July 28, 2022, https://www.newscientist.com/article/2330866-deepminds-protein-folding-ai-cracks-biologys-biggest-problem/.

5. Sergey Nurk, Sergey Koren, Arang Rhie, Mikko Rautiainen, Andrey V. Bzikadze, Alla Mikheenko, Mitchell R. Vollger, Nicolas Altemose, Lev Uralsky, Ariel Gershman, et al., "The Complete Sequence of the Human Genome," *Science* 376, no. 6588 (April 1, 2022): 44–53, https://www.science.org/doi/10.1126/science.abj6987; Jennifer E. Rood and Aviv Regev, "The Legacy of the Human Genome Project," *Science* 373, no. 6562 (September 24, 2021): 1442–43, https://www.science.org/doi/10.1126/science.abl5403.

6. H. Holden Thorp, "Editorial: Proteins, Proteins Everywhere," *Science* 374, no. 6574 (December 17, 2021): 1415, https://www.science.org/doi/10.1126/science.abn5795.

7. Service, "Protein Structures for All."

8. Daniel J. Boorstin, *The Creators* (New York: Vintage Books, 1993); Daniel J. Boorstin, *The Discoverers* (London: J. M. Dent, 1984); Walter Isaacson, *Benjamin Franklin: An American Life* (New York: Simon and Schuster, 2003); Walter Isaacson, *The Innovators: How a Group of Hackers, Geniuses, and Geeks Created the Digital Revolution* (New York: Simon and Schuster, 2014); Walter Issacson, *Steve Jobs* (New York: Simon and Schuster, 2011).

9. Benjamin Radford, "Jan. 6 Investigation Testimony Reveals Conspiracy-Riddled Trump White House," *Skeptical Inquirer* 44, no. 5 (September/October 2022): 5–6.

10. Steven Pinker, *Enlightenment Now: The Case for Reason, Science, Humanism, and Progress* (New York: Viking, 2018); Hans Rosling, with Ola Rosling and Anna Rosling Rönnlund, *Factfulness: Ten Reasons We're Wrong about the World—and Why Things Are Better Than You Think* (New York: Flatiron Books, 2018).

11. Rosling, Rosling, and Rönnlund, *Factfulness*, 15.

12. Hans Rosling and Ola Rosling, "A Fact-Based Worldview through Animated Data" (presentation, "ESCape to Clarity!" Fifteenth European Skeptics Congress, Stockholm, Sweden, August 23–25, 2013).

13. Kendrick Frazier, "Three Days of Science and Skepticism in Stockholm," *Skeptical Inquirer* 38, no. 1 (January/February 2014): 5–7.

14. Rosling, Rosling, and Rönnlund, *Factfulness*, 8.

15. Rosling, Rosling, and Rönnlund, *Factfulness*, 8.

16. Rosling, Rosling, and Rönnlund, *Factfulness*, 8.

17. Rosling, Rosling, and Rönnlund, *Factfulness*, 8.

18. Daniel Kahneman, *Thinking, Fast and Slow* (New York: Farrar, Straus, and Giroux, 2011).

19. Rosling, Rosling, Rönnlund, *Factfulness*, 8.

20. Pinker, *Enlightenment Now*.

21. Pinker, *Enlightenment Now*.

22. Pinker, *Enlightenment Now*, 4–5.

23. Pinker, *Enlightenment Now*.

24. Steven Pinker, "Progressophobia: Why Things Are Better Than You Think They Are," *Skeptical Inquirer* 43, no. 3 (May/June 2018): 26–35.

25. Pinker, *Enlightenment Now*.

26. Pinker, *Enlightenment Now*.

27. Joshua Rothman, "The Big Question: Is the World Getting Better or Worse?" *New Yorker*, July 28, 2018, 26–32.

28. Steven Pinker, "Enlightenment Wars: Some Reflections on 'Enlightenment Now,' One Year Later," *Quillette*, January 14, 2019, https://quillette.com/2019/01/14/enlightenment-wars-some-reflections-on-enlightenment-now-one-year-later/.

29. Pinker, "Enlightenment Wars."

30. Pinker, "Enlightenment Wars."

31. Pinker, "Enlightenment Wars."

32. See Robyn Blumner, "I Choose Optimism," *Free Inquiry* 42, no. 5 (August/September 2022): 4; Benjamin Radford, *America the Fearful: Media and the Marketing of National Panics* (Jefferson, NC: McFarland, 2022).

BIBLIOGRAPHY

Abell, George O., and Barry Singer, eds. *Science and the Paranormal: The Existence of the Supernatural.* New York: Scribner's, 1981.

Abbott, B. P., R. Abbott, T. D. Abbott, M. R. Abernathy, F. Acernese, K. Ackley, C. Adams, T. Adams, R. X. Adhikari, V. B. Adya, et al. "Observation of Gravitational Waves from a Binary Black Hole Merger." *Physical Review Letters* 116, no. 061102 (February 11, 2016). https://journals.aps.org/prl/pdf/10.1103/PhysRevLett.116.061102.

Achenbach, Joel. "Big Bang Backlash: BICEP2 Discovery of Gravity Waves Questioned by Cosmologists." *Washington Post*, May 17, 2014, A1–11. https://www.washingtonpost.com/national/health-science/big-bang-backlash-bicep2-discovery-of-gravity-waves-questioned-by-cosmologists/2014/05/16/e575b2fc-db07-11e3-bda1-9b46b2066796_story.html.

———. "Enigmatic Skull Suggests Our Human Species Reached Europe 210,000 Years Ago." *Washington Post*, July 10, 2019. https://www.washingtonpost.com/science/2019/07/10/enigmatic-skull-suggests-our-human-species-reached-europe-years-ago/.

Ajmera, Maya. "Conversations with Maya." Interview by Thomas Rosenbaum. *Science News* 195, no. 6 (March 30, 2019): 28–29.

Alcock, James E. *Belief: What It Means to Believe and Why Our Convictions Are So Compelling.* Amherst, NY: Prometheus Books, 2018.

———. "Extrasensory Perception." In *The Encyclopedia of the Paranormal*, edited by Gordon Stein, 241–54. Amherst, NY: Prometheus Books, 1996.

———. *Parapsychology: Science or Magic? A Psychological Perspective.* New York: Pergamon Press, 1981.

———. *Science and Supernature: A Critical Appraisal of Parapsychology*. Amherst, NY: Prometheus Books, 1990.

Alcock, James, Jean Burns, and Anthony Freeman. *Psi Wars: Getting to Grips with the Paranormal.* Exeter, UK: Imprint Academic, 2003.

Al-Khalili, Jim. *The Joy of Science.* Princeton, NJ: Princeton University Press, 2022.

Alvarez, Luis W., Walter Alvarez, Frank Asaro, and Helen V. Michel. "Extraterrestrial Cause for the Cretaceous-Tertiary Extinction." *Science* 208, no. 4448 (June 6, 1980): 1095–1108. https://doi.org/10.1126/science.208.4448.1095.

Amundson, Ron. "Follow-Up: Watson and the 'Hundredth Monkey Phenomenon.'" *Skeptical Inquirer* 11, no. 3 (Spring 1987): 303–4.

———. "The Hundredth Monkey Phenomenon." *Skeptical Inquirer* 9, no. 4 (Summer 1985): 348–56.

Amunts, Alexey, Alan Brown, Xiao-Chen Bai, Jose L. Llácer, Tanweer Hussain, Paul Emsley, Fei Long, Garib Murshudov, Sjors H. W. Scheres, and V. Ramakrishnan. "Structure of the Yeast Mitochondrial Large Ribosomal Subunit." *Science* 343, no. 6178 (March 28, 2014): 1485–89. https://www.science.org/doi/10.1126/science.1249410.

Andersen, Kurt. *Fantasyland: How America Went Haywire: A 500-Year History.* New York: Random House, 2017.

Angela, Piero. "The World of Galileo Galilei." Presentation at the Fifth World Skeptics Congress, Abano Terme, Italy, October 8–10, 2004.

Annenberg Public Policy Center. "Millions Embrace Covid-19 Misinformation, Which Is Linked to Hesitancy on Vaccination and Boosters." University of Pennsylvania. December 17, 2021. https://www.annenbergpublicpolicycenter.org/millions-embrace-covid-19-misinformation-which-is-linked-to-hesitancy-on-vaccination-and-boosters/.

Asimov, Isaac. "Asimov's Corollary." *Skeptical Inquirer* 3, no. 3 (Spring 1979): 58–67. https://skepticalinquirer.org/1979/04/asimovs_corollary/.

———. *Past, Present, and Future*. Buffalo, NY: Prometheus Books, 1987.

———. "The Perennial Fringe." *Skeptical Inquirer* 10, no. 3 (Spring 1986): 212–14.

———. *The Roving Mind.* New ed. Amherst, NY: Prometheus Books, 1997.

Asimov, Isaac, and Jason A. Shulman, eds. *Isaac Asimov's Book of Science and Nature Quotations*. New York: Weidenfeld and Nicolson, 1986.

Atwood, Kimball. "The Ongoing Problem with the National Center for Complementary and Alternative Medicine." *Skeptical Inquirer* 27, no. 5 (September/October 2003).

Baek, Minkyung, Frank Dimaio, Ivan Anishchenko, Justas Dauparas, Sergey Ovchinnikov, Gyu Rie Lee, Jue Wang, Qian Cong, Lisa N. Kinch, R. Dustin

Schaeffer, et al. "Accurate Prediction of Protein Structures and Interactions Using a Three-Track Neural Network." *Science* 373, no. 6557 (August 19, 2021): 871–76. https://doi.org/10.1126/science.abj8754.

Baggini, Julian. *The Edge of Reason: A Rational Skeptic in an Irrational World.* New Haven, CT: Yale University Press, 2016.

Bais, Sander. *In Praise of Science: Curiosity, Understanding, and Progress.* Cambridge, MA: MIT Press, 2010.

Basterfield, Candice, Scott O. Lilienfeld, Shauna M. Bowes, and Thomas H. Costello. "The Nobel Disease: When Intelligence Fails to Protect against Rationality." *Skeptical Inquirer* 44, no. 3 (May/June 2020): 32–37.

Bauer, Joseph A. "A Surgeon's View of the 'Alien Autopsy.'" *Skeptical Inquirer* 20, no. 1 (January/February 1996).

Bergstrom, Carl T., and Jevin D. West. *Calling Bullshit: The Art of Skepticism in a Data-Driven World.* New York: Random House, 2020.

Berman, Jonathan M. *Anti-vaxxers: How to Challenge a Misinformed Movement.* Cambridge, MA: MIT Press, 2020.

Berthold, Richard M. *Dare to Struggle: The History and Society of Ancient Greece.* N.p.: iUniverse, 2009.

———. Lecture at a meeting of New Mexicans for Science and Reason, April 11, 2007.

Bertman, Stephen. *The Genesis of Science: The Story of Greek Imagination.* Amherst, NY: Prometheus Books, 2010.

BICEP2/Keck and Planck Collaboration. "Joint Analysis of BICEP2/*Keck Array* and *Planck* Data." *Physical Review Letters* 114, no. 101301 (March 9, 2015). https://journals.aps.org/prl/abstract/10.1103/PhysRevLett.114.101301.

Blaskiewicz, Robert, and Mike Jursulic. "Arthur C. Cramp: The Quackbuster Who Professionalized American Medicine." *Skeptical Inquirer* 42, no. 6 (November/December 2018): 45–50.

Bloom, Allen. *The Closing of the American Mind.* New York: Simon and Schuster, 1987.

Blumner, Robyn. "I Choose Optimism." *Free Inquiry* 42, no. 5 (August/September 2022):4.

———. "Postcards from Reality: Why Full of Bull? Here's Why." *Skeptical Inquirer* 45, no. 6 (November/December 2021): 14–15.

Bok, Bart J. "A Critical Look at Astrology." *Humanist* 35, no. 5 (September/October 1975): 6–9.

Bok, Bart J., and Lawrence E. Jerome. *Objections to Astrology: 192 Leading Scientists, Including 19 Nobel Prize Winners, Disavow Astrology.* Amherst, NY: Prometheus Books, 1975.

Bok, Bart J., Lawrence E. Jerome, Paul Kurtz, et al. "Objections to Astrology: A Statement by 186 Leading Scientists." *Humanist* 35, no. 5 (September/October 1975): 4–6.

Boorstin, Daniel J. *The Creators*. New York: Vintage Books, 1993.

———. *The Discoverers*. London: J. M. Dent, 1984.

Braga, Brannon, Bill Pope, and Ann Druyan, dirs. *Cosmos: A Spacetime Odyssey*. Written by Ann Druyan and Steven Soter, aired March 9–June 8, 2014, on National Geographic Channel.

Broch, Henri. *Exposed! Ouija, Firewalking, and Other Gibberish*. Baltimore, MD: Johns Hopkins University Press, 2009.

Bronowski, J. *The Ascent of Man*. London: BBC, 1973.

———. *The Common Sense of Science*. Cambridge, MA: Harvard University Press, 1976.

———. *The Origins of Knowledge and Imagination*. New Haven, CT: Yale University Press, 1978.

Bunge, Mario. *Between Two Worlds: Memoirs of a Philosopher-Scientist*. New York: Springer Berlin Heidelberg, 2016.

———. "The Dematerialization Crusade." *Skeptical Inquirer* 43, no. 2 (March/April 2019): 56–58.

———. *From a Scientific Point of View: Reasoning and Evidence Beat Improvisation across Fields*. Newcastle upon Tyne, UK: Cambridge Scholars, 2018.

———. "The Philosophy behind Pseudoscience." *Skeptical Inquirer* 30, no. 4 (July/August 2006): 29–37.

———. "The Scientist's Skepticism." *Skeptical Inquirer* 44, no. 6 (November/December 2020): 57–59.

———. "What Is Pseudoscience?" *Skeptical Inquirer* 9, no. 1 (Fall 1984): 36–46.

Burtnyk, Kimberly. "LIGO Celebrates First Anniversary of Historic Gravitational Wave Detection!" Laser Interferometer Gravitational-Wave Observatory. September 14, 2006. https://www.ligo.caltech.edu/news/ligo20160914.

Byler, David, and Yan Wu. "Opinion: Will You Fall into the Conspiracy Theory Rabbit Hole? Take Our Quiz and Find Out." *Washington Post*, October 6, 2021. https://www.washingtonpost.com/opinions/interactive/2021/conspiracy-theory-quiz/.

Cahn, J. P. "The Flying Saucers and the Mysterious Little Men." *True*, September 1952, 17–19, 102–12.

Callaway, Ewan. "'It Will Change Everything': DeepMind's AI Makes Gigantic Leap in Solving Protein Structures." *Nature* 588, no. 7837 (December 10, 2020): 203–4.

Cartlidge, Edwin. "Have Physicists Seen the Dying Flash of Dark Matter?" *Science*, 350, no. 6256 (October 2, 2015): 20. https://www.science.org/doi/full/10.1126/science.350.6256.20.

Castelvecchi, Davide. "Black Hole at the Centre of Our Galaxy Imaged for the First Time." *Nature* 605 (May 12, 2022): 403–4. https://doi.org/10.1038/d41586-022-01320-y.

Caulfield, Timothy. *Is Gwyneth Paltrow Wrong about Everything? How the Famous Sell Us Elixirs of Health, Beauty and Happiness.* Boston: Beacon Press, 2015.

"Center for Inquiry, Richard Dawkins Foundation Merging; Robyn Blumner CFI's New CEO." *Skeptical Inquirer* 40, no. 3 (May/June 2016): 5–6.

Chamovitz, Daniel. *What a Plant Knows: A Field Guide to the Senses*. New York: Scientific American Books, 2012.

Charpak, Georges, and Henri Broch. *Debunked! ESP, Telekinesis, and Other Pseudosciences.* Baltimore, MD: Johns Hopkins University Press, 2004.

Cho, Adrian. "Blockbuster Claim Could Collapse in a Cloud of Dust." *Science* 344, no. 6186 (May 23, 2014): 790. https://www.science.org/doi/10.1126/science.344.6186.790.

Christian, James L., ed. *Extraterrestrial Intelligence: The First Encounter*. Amherst, NY: Prometheus Books, 1976.

Clairborne, Robert. *Climate, Man, and History*. New York: Norton, 1970.

Clarke, Arthur C. *Profiles of the Future: An Inquiry into the Limits of the Possible.* Rev. ed. New York: Harper and Row, 1973.

Clery, Daniel. "Webb Reveals Early Universe's Galactic Bounty." *Science* 377, no. 6607 (August 12, 2022): 700–701. https://doi.org/10.1126/science.ade3375.

———. "Webb Telescope Reveals Unpredicted Bounty of Bright Galaxies in Early Universe." *Science*. August 9, 2022. https://www.science.org/content/article/webb-telescope-reveals-unpredicted-bounty-bright-galaxies-early-universe.

CO2.Earth. "Earth's CO_2 Home Page." Accessed April 15, 2023. https://www.co2.earth.

"Committee for Skeptical Inquiry Names Ten New Fellows." *Skeptical Inquirer* 45, no. 1 (January/February 2021).

Cooper, Alan, Chris S. M. Turney, Jonathan Palmer, Alan Hogg, Matt McGlone, Janet Wilmshurst, Andrew M. Lorrey, et al. "A Global Environmental Crisis 42,000 Years Ago." *Science* 371, no. 6531 (February 19, 2021): 817. https://doi.org/10.1126/science.abb8677.

———. "Response to Comment on 'A Global Environmental Crisis 42,000 Years Ago.'" *Science* 374, no. 6570 (November 18, 2021). https://doi.org/10.1126/science.abh3655.

———. "Response to Comment on 'A Global Environmental Crisis 42,000 Years Ago.'" *Science* 374, no. 6570 (November 18, 2021), https://doi.org/10.1126/science.abi9756.

Couzin-Frankel, Jennifer. "Cancer Immunotherapy." *Science* 342, no. 6165 (December 20, 2013): 1432–33. https://www.science.org/doi/10.1126/science.342.6165.1432.

Cowen, Ron. "Full-Galaxy Dust Map Muddles Search for Gravitational Waves." *Nature* (2014). https://doi.org/10.1038/nature.2014.15975.

———. "Gravitational Waves Discovery Now Officially Dead." *Nature* (January 30, 2015). https://doi.org/10.1038/nature.2015.16830.

Crockett, Christopher. "Gravitational Waves Unmask Universe Just after Big Bang." *Science News* 185, no. 7 (April 5, 2014). https://www.sciencenews.org/article/gravitational-waves-unmask-universe-just-after-big-bang.

Damon, P. E., D. J. Donahue, B. H. Gore, A. L. Hatheway, A. J. T. Jull, T. W. Linick, P. J. Sercel, et al. "Radiocarbon Dating of the Shroud of Turin." *Nature* 337, no. 6208 (February 16, 1989): 611–15.

Dawkins, Richard. *The Extended Selfish Gene*. Oxford, UK: Oxford University Press, 2016.

———. *The Greatest Show on Earth: The Evidence for Evolution.* New York: Free Press, 2010.

———. *The Magic of Reality: How We Know What's Really True.* New York: Free Press, 2011.

———. "Science, Delusion, and the Appetite for Wonder." *Edge*. December 29, 1996. http://edge.org/conversation/science-delusion-and-the-appetite-for-wonder.

———. "Science: The Gold Standard of Truth." *Skeptical Inquirer* 45, no. 2 (March/April 2021): 38–40.

———. *Science in the Soul: Selected Writings of a Passionate Rationalist.* New York: Random House, 2017.

———. *Unweaving the Rainbow: Science, Delusion, and the Appetite for Wonder.* Boston: Houghton Mifflin, 1998.

DeepMind and EMBL-EBI. *AlphaFold Protein Structure Database*. Accessed April 16, 2023. https://alphafold.ebi.ac.uk.

DeMarban, Alex. "Storms Pummel Bering Sea Islands after 'Crazy' Ice Melt-Off." *Anchorage Daily News*, February 28, 2018. https://www.adn.com/alaska-news/rural-alaska/2018/02/27/warm-storms-pummel-bering-sea-leading-to-crazy-ice-melt-off/#:~:text=A%20swath%20of%20sea%20ice,17%20years%20ago%2C%20scientists%20say.

De Sanctis, M. C., E. Ammannito, A. Raponi, A. Frigeri, M. Ferrari, F. G. Carrozzo, M. Ciarniello, M. Formisano, B. Rousseau, F. Tosi, et al. "Fresh

Emplacement of Hydrated Sodium Chloride on Ceres from Ascending Salty Fluids." *Nature Astronomy* 4 (August 10, 2020): 786–93. https://www.nature.com/articles/s41550-020-1138-8.

Druyan, Ann. *Cosmos: Possible Worlds*. Washington, DC: National Geographic, 2020.

Durant, Will, and Ariel Durant. *The Age of Reason Begins*. Pt. 7, *The Story of Civilization*. New York: Simon and Schuster, 1961.

Earth Observatory. "Roger Revelle (1909–1991)." June 19, 2000. https://earthobservatory.nasa.gov/Features/Revelle/revelle_3.php.

Eberhardt, Jennifer L. *Biased: Uncovering the Hidden Prejudice That Shapes What We See, Think, and Do*. New York: Viking, 2019.

Edis, Taner. *Weirdness: What Fake Science and the Paranormal Tell Us about the Nature of the Universe*. Durham, NC: Pitchstone, 2021.

Einstein, Albert. "Imagination is more important than knowledge." Quote Investigator. January 1, 2013. http://quoteinvestigator.com/2013/01/01 einstein-imagination/.

Eisenberg, David M., Ronald C. Kessler, Cindy Foster, Frances E. Norlock, David R. Calkins, and Thomas L. Delbanco. "Unconventional Medicine in the United States." *New England Journal of Medicine* 328, no. 4 (January 28, 1993): 246–52. http://www.nejm.org/doi/pdf/10.1056/NEJM199301283280406.

Ernst, Edzard. "Alternative Medicine Is a Playground for Apologists." *Skeptical Inquirer* 40, no. 5 (September/October 2016).

———. *Charles, the Alternative Prince: An Unauthorised Biography*. Exeter, UK: Imprint Academic, 2022.

———. *Don't Believe What You Think: Arguments for and against SCAM*. London: Imprint Academic, 2020.

———. *Edzard Ernst* (blog). Accessed April 11, 2023. https://edzardernst.com.

———. *SCAM: So-Called Alternative Medicine*. London: Imprint Academic, 2018.

———. *So-Called Alternative Medicine (SCAM) for Cancer*. Cham, Switzerland: Springer, 2021.

Ernst, Edzard, Max H. Pittler, Clare Stevinson, and Adrian White, eds. *The Desktop Guide to Complementary and Alternative Medicine*. Edinburgh, UK: Mosby, 2001.

Ernst, Edzard, and Kevin Smith. *More Harm than Good? The Moral Maze of Complementary and Alternative Medicine*. Cham, Switzerland: Springer, 2018.

European Space Agency. "Planck: Gravitational Waves Remain Elusive." January 30, 2015. http://sci.esa.int/planck/55362-planck-gravitational-waves-remain-elusive.

Event Horizon Telescope Collaboration. "First M87 Event Horizon Telescope Results. I. The Shadow of the Supermassive Black Hole." *Astrophysical Journal Letters* 875, no. 1 (April 10, 2019). https://iopscience.iop.org/article/10.3847/2041-8213/ab0ec7.

Fagan, Brian. *The Little Ice Age: How Climate Made History, 1300–1850.* New York: Basic Books, 2000.

———. *The Long Summer: How Climate Changed Civilization.* New York: Basic Books, 2004.

Fahy, Declan. *The New Celebrity Scientists: Out of the Lab and into the Limelight.* Lanham, MD: Rowman and Littlefield, 2015.

Falcke, Heino. *Light in the Darkness: Black Holes, the Universe, and Us.* San Francisco: HarperOne, 2021.

Farha, Bryan., ed. *Paranormal Claims: A Critical Analysis.* Lanham, MD: University Press of America, 2007.

———, ed. *Pseudoscience and Deception: The Smoke and Mirrors of Paranormal Claims.* Lanham, MD: University Press of America, 2014.

Farley, Tim. *What's the Harm?* (blog). Accessed April 7, 2023. http://whatstheharm.net/index.html.

Feynman, Richard. "Solutions of Maxwell's Equations in Free Space." In *The Feynman Lectures on Physics*, by Richard P. Feynman, Robert B. Leighton, and Matthew Sands. Vol. 2, *Mainly Electromagnetism and Matter.* Redwood City, CA: Addison-Wesley, 1989. https://www.feynmanlectures.caltech.edu/.

———. *What Do* You *Care What Other People Think?* New York: Norton, 1988.

Frankfurt, Harry G. *On Bullshit.* Princeton, NJ: Princeton University Press, 2005.

Frazier, Kendrick. "Challenging Pseudoscience." *Science News* 109, no. 22 (May 29, 1976).

———. "Committee for the Scientific Investigation of Claims of the Paranormal (CSICOP)." In *The Encyclopedia of the Paranormal*, edited by Gordon Stein (168–81). Amherst, NY: Prometheus Books, 1996. https://skepticalinquirer.org/history-of-csicop/.

———. "CSICOP Timeline: A Capsule History in 85 Easy Steps." *Skeptical Inquirer* 25, no. 3 (May/June 2001). http://www.skepticalinquirer.org/si/show/csicop_timeline.

———. "Editorial: Explanatory Frameworks and Investigative Exposés." *Skeptical Inquirer* 30, no. 4 (July/August 2006): 4.

———, ed. *Encounters with the Paranormal: Science, Knowledge, and Belief.* Amherst, NY: Prometheus Books, 1998.

———. "First Contact: The News Event and the Human Response." In *Extraterrestrial Intelligence: The First Encounter*, edited by James Christian (73–88). Amherst, NY: Prometheus Books, 1976.

———, ed. *The Hundredth Monkey and Other Paradigms of the Paranormal: A* Skeptical Inquirer *Collection.* Buffalo, NY: Prometheus Books, 1991.

———. "In Troubled Times, This Is What We Do." *Skeptical Inquirer* 42, no. 2 (March/April 2018): 14–15.

———, ed. "Issues in Science and Skepticism." Special issue, *Skeptical Inquirer* 40, no. 5 (September/October 2016).

———. "It's CSI Now, Not CSICOP." *Skeptical Inquirer* 31, no. 1 (January/February 2007): 5–6.

———. "Neil deGrasse Tyson on Space, Reality, Pop Culture, Skeptics, and Atheists." *Skeptical Inquirer* 45, no. 6 (September/October 2021):6–7.

———, ed. "Odysseys in Scientific Skepticism." Special issue, *Skeptical Inquirer* 40, no. 6 (November/December 2016).

———. "Organized Skepticism: Four Decades . . . and Today." *Skeptical Inquirer* 39, no. 2 (March/April 2015): 14–18.

———. *Our Turbulent Sun: The Scientific Quest for a New Understanding of Our Sun and Its Surprising Irregularities, Variations, and Effects*. Englewood Cliffs, NJ: Prentice-Hall, 1982.

———, ed. *Paranormal Borderlands of Science*. Buffalo, NY: Prometheus Books, 1981.

———. "Prometheus Unbound: Publisher of Skeptic, Freethought Books Enters a New Phase." *Skeptical Inquirer* 44, no. 3 (May/June 2020): 6–9.

———. "The Roswell Incident at 70: Facts, Not Myths." *Skeptical Inquirer* 41, no. 6 (November/December 2017).

———. "Science and the Parascience Cults." *Science News* 109 (May 29, 1977): 346–48, 350.

———, ed. *Science Confronts the Paranormal.* Buffalo, NY: Prometheus Books, 1986.

———, ed. S*cience under Siege: Defending Science, Exposing Pseudoscience.* Amherst, NY: Prometheus Books, 2009.

———. "The State of Our Nation (From the Editor)." *Skeptical Inquirer* 44, no. 5 (September/October 2020): 4.

———. "Three Days of Science and Skepticism in Stockholm." *Skeptical Inquirer* 38, no. 1 (January/February 2014): 5–7.

———. "UFOs, Horoscopes, Bigfoot, Psychics, and Other Nonsense." *Smithsonian* 8, no. 12 (March 1978): 55–60.

———. "The Velikovsky Affair: The Distortions Continue." *Skeptical Inquirer* 5, no. 1 (Fall 1980).

———. *The Violent Face of Nature: Severe Phenomena and Natural Disasters*. New York: Morrow, 1979.

———. "What Exxon Knew about Arctic Warming Twenty-Five Years Ago." *Skeptical Inquirer* 40, no. 1 (January/February 2016): 5–6.

———. "Why We Do This: Revisiting the Higher Values of Skeptical Inquiry." *Skeptical Inquirer* 37, no. 6 (November/December 2013): 13.

———. "The Winter of Our Discontent." *Skeptical Inquirer* 34, no. 3 (May/June 2010): 4.

———. "You Can't Fit What We Skeptics Do into a Neat Box." *Skeptical Inquirer* 43, no. 2 (March/April 2019): 22–23.

Frazier, Kendrick, Barry Karr, and Joe Nickell, eds. *The UFO Invasion: The Roswell Incident, Alien Abductions, and Government Coverups*. Amherst, NY: Prometheus Books, 1997.

French, Christopher, and Anna Stone. *Anomalistic Psychology: Exploring Paranormal Belief and Experience*. London: Palgrave Macmillan, 2014.

Friedlander, Michael W. *At the Fringes of Science*. Boulder, CO: Westview Press, 1995.

Galef, Julia. *The Scout Mindset: Why Some People See Things Clearly and Others Don't*. New York: Portfolio, 2021.

Galston, Arthur W., and Clifford L. Slayman. "The Not-So-Secret Life of Plants." *American Scientist* 67 (May 1979): 337–44.

Gardner, Martin. *Did Adam and Eve Have Navels? Discourses on Reflexology, Numerology, Urine Therapy, and Other Dubious Subjects*. New York: W. W. Norton, 1957.

———. *Fads and Fallacies in the Name of Science*. New York: Dover, 2000.

———. *In the Name of Science*. New York: Putnam, 1952.

———. *The New Age: Notes of a Fringe Watcher*. Buffalo, NY: Prometheus Books, 1988.

———. "Notes of a Psi-Watcher: Lessons of a Landmark PK Hoax." *Skeptical Inquirer*. 7, no. 4 (Summer 1983): 16–19.

———. "Oprah Winfrey: Bright (but Gullible) Billionaire." *Skeptical Inquirer* 34, no. 2 (March/April 2010): 55–56.

———. *Science: Good, Bad, and Bogus*. Amherst, NY: Prometheus Books, 1981.

Gillon, Michaël, Emmanuël Jehin, Susan M. Lederer, Laetitia Delrez, Julien de Wit, Artem Burdanov, Valérie Van Grootel, Adam J. Burgasser, Amaury H. M. J. Triaud, Cyrielle Opitom, et al. "Temperate Earth-Sized Planets Transiting a Nearby Ultracool Dwarf Star." *Nature* 533 (February 23, 2016): 221–24.

Goop. "What's Goop?" Accessed April 11, 2023. https://goop.com/whats-goop/.

Gordin, Michael D. *On the Fringe: Where Science Meets Pseudoscience*. New York: Oxford University Press, 2021.

———. *The Pseudoscience Wars: Immanuel Velikovsky and the Birth of the Modern Fringe*. Chicago: University of Chicago Press, 2012.

Gore, Al. *An Inconvenient Truth: The Crisis of Global Warming*. New York: Viking, 2007.

Gorski, David. "Antivaxxers Rejoice at the 'Silver Lining' from the Pandemic, Spillover of Distrust of COVID-19 Vaccines to All Vaccines." Science-Based Medicine. August 15, 2022. https://sciencebasedmedicine.org/antivaxxers-rejoice-at-the-silver-lining-from-the-pandemic-spillover/.

Gould, Stephen Jay. *Dinosaur in a Haystack: Reflections in Natural History*. New York: W. W. Norton, 1995.

———. *The Mismeasure of Man*. New York: W. W. Norton, 1981.

Grim, Patrick, ed. *Philosophy of Science and the Occult*. 2nd ed. Albany: State University of New York Press, 1990.

Grimes, David Robert. *Good Thinking: Why Flawed Logic Puts Us All at Risk and How Critical Thinking Can Save the World*. New York: Experiment, 2021.

———. "Schrödinger's Bin Laden: The Irrational World of Motivated Reasoning." *Skeptical Inquirer* 46, no. 1 (January/February 2022): 34–39.

Gross, Paul R., and Norman Levitt. *Higher Superstition: The Academic Left and Its Quarrels with Science*. Baltimore, MD: Johns Hopkins University Press, 1994.

Grossman, Wendy M., and Christopher C. French, eds. *Why Statues Weep: The Best of* The Skeptic. London: Philosophy Press, 2010.

Guggenheim, Davis, dir. *An Inconvenient Truth*. Written by Al Gore. Beverly Hills, CA: Lawrence Bender Productions, 2006.

Haack, Susan. *Defending Science—within Reason: Between Scientism and Cynicism*. Amherst, NY: Prometheus Books, 2003.

———. *Manifesto of a Passionate Moderate: Unfashionable Essays*. Chicago: University of Chicago Press, 1998.

———. *Philosophy of Logics*. Cambridge, UK: Cambridge University Press, 1978.

———. *Putting Philosophy to Work: Inquiry and Its Place in Culture: Essays on Science, Religion, Law, Literature, and Life*. Amherst, NY: Prometheus Books, 2013.

Hall, Harriet. "The Care and Feeding of the Vagina." *Skeptical Inquirer* 42, no. 5 (September/October 2018): 28–29.

———. "Is Acupuncture Winning?" *Skeptical Inquirer* 43, no. 1 (January/February 2019): 19–21.

———. "The Prince of Alternative Medicine: Review of Edzard Ernst, *Charles, the Alternative Prince*." *Skeptical Inquirer* 46, no. 3 (May/June 2022): 59–61.

———. "Science Envy in Alternative Medicine." *Skeptical Inquirer* 43, no. 4 (July/August 2019): 21–23.

———. "Why We Need Science." *Skeptical Inquirer* 44, no. 6 (November/December 2020): 40–42.

———. "Wither Chiropractic." *Skeptical Inquirer* 43, no. 6 (November/December 2020): 24–26.

———. *Women Aren't Supposed to Fly: The Memoirs of a Female Flight Surgeon.* N.p.: iUniverse, 2008.

Hansson, Sven Ove. "Defining Pseudoscience and Science." In *Philosophy of Pseudoscience: Reconsideration of the Demarcation Problem*, edited by Massimo Pigliucci and Maarten Boudry. Chicago: University of Chicago Press, 2013.

Hargittai, Istvan. *Drive and Curiosity: What Fuels the Passion for Science.* Amherst, NY: Prometheus Books, 2011.

Harrison, Guy. *50 Popular Beliefs That People Think Are True*. Amherst, NY: Prometheus Books, 2012.

———. *Good Thinking: What You Need to Know to Be Smarter, Safer, Wealthier, and Wiser.* Amherst, NY: Prometheus Books, 2015.

———. "How to Repair the American Mind." *Skeptical Inquirer* 43, no. 3 (May/June 2021): 31–34.

———. *Think: Why You Should Question Everything*. Amherst, NY: Prometheus Books, 2013.

Harvati, Katerina, Carolin Röding, Abel M. Bosman, Fotios A. Karakostis, Rainer Grün, Chris Stringer, Panagiotis Karkanas, Nicholas C. Thompson, Vassilis Koutoulidis, Lia A. Moulopoulos, et al. "Apidima Cave Fossils Provide Earliest Evidence of *Homo sapiens* in Eurasia." *Nature* 571 (July 10, 2019): 500–504. https://www.nature.com/articles/s41586-019-1376-z.

Hawks, John. "Comment on 'A Global Environmental Crisis 42,000 Years Ago." *Science* 374, no. 6570 (November 18, 2021). https://doi.org/10.1126/science.abh1878.

Hays, J. D., John Imbrie, and N. J. Shackleton. "Variations in the Earth's Orbit: Pacemaker of the Ice Ages." *Science* 194, no. 4270 (December 10, 1976): 1121–32.

Helfand, David J. *A Survival Guide to the Misinformation Age: Scientific Habits of Mind.* New York: Columbia University Press, 2016.

Henson, Robert. *The Rough Guide to Climate Change.* London: Rough Guides, 2008.

Hilgartner, Stephen, J. Benjamin Hurlbut, and Sheila Jasanoff. "Was 'Science' on the Ballot? (Policy Forum)." *Science* 371, no. 6532 (February 26, 2021): 893–94.

Hill, Carol, Gregg Davidson, Tim Helble, and Wayne Ranney, eds. *The Grand Canyon, Monument to an Ancient Earth: Can Noah's Flood Explain the Grand Canyon?* Tulsa, OK: Kregel, 2016.

Hill, Sharon. *Scientifical Americans: The Culture of Amateur Paranormal Researchers*. Jefferson, NC: McFarland, 2017.

Hines, Terence. *Pseudoscience and the Paranormal*. 2nd ed. Amherst, NY: Prometheus Books, 2003.

Hofstadter, Douglas. "World Views in Collision: *The Skeptical Inquirer* versus the *National Enquirer*." In *Metamagical Themas: Questing for the Essence of Mind and Pattern*. New York: Basic Books, 1985.

Holton, Gerald. *Science and Anti-Science*. Cambridge, MA: Harvard University Press, 1993.

Horowitz, Kenneth A., Donald G. Lewis, and Edgar L. Gasteiger. "Plant 'Primary Perception': Electrophysiological Unresponsiveness to Brine Shrimp Killing." *Science* 189 (1976): 478–80.

Hotez, Peter, J. *Preventing the Next Pandemic: Vaccine Diplomacy in a Time of Anti-Science*. Baltimore, MD: Johns Hopkins University Press, 2021.

Houdini, Harry. *Miracle Mongers and Their Methods: A Complete Exposé*. Amherst, NY: Prometheus Books, 1981.

Hupp, Stephen, ed. *Pseudoscience in Child and Adolescent Therapy: A Skeptic's Guide*. Cambridge, UK: Cambridge University Press, 2019.

Hyman, Ray. "Commentary on John P.A. Ionnidis's 'Why Most Published Research Findings Are False.'" *Skeptical Inquirer* 30, no. 2 (March/April 2006): 17–18.

———. *The Elusive Quarry: A Scientific Appraisal of Psychical Research*. Amherst, NY: Prometheus Books, 1989.

Imbrie, John, and Katherine Palmer Imbrie. *Ice Ages: Solving the Mystery*. Short Hills, NJ: Enslow, 1979.

Ionnidis, John P. A. "Why Most Published Research Findings Are False." *PLoS Medicine* 2, no. 8 (August 2005): 696–701.

Isaacson, Walter. *Benjamin Franklin: An American Life*. New York: Simon and Schuster, 2003.

———. *The Innovators: How a Group of Hackers, Geniuses, and Geeks Created the Digital Revolution*. New York: Simon and Schuster, 2014.

———. *Steve Jobs*. New York: Simon and Schuster, 2011.

Jacoby, Susan. *The Age of American Unreason*. Rev. ed. New York: Vintage Books, 2009.

Jumper, John, Richard Evans, Alexander Pritzel, Tim Green, Michael Figurnov, Olaf Ronneberger, Kathryn Tunyasuvunakool, Russ Bates, Augustin Žídek, Anna Potapenko, et al. "Highly Accurate Protein Structure Prediction with AlphaFold." *Nature* 596 (August 26, 2021): 583–89. https://doi.org/10.1038/s41586-021-03819-2.

Kahneman, Daniel. *Thinking, Fast and Slow*. New York: Farrar, Straus, and Giroux, 2011.

Kahn-Harris, Keith. *Denial: The Unspeakable Truth*. Mirefoot, UK: Notting Hill Editions, 2018.

Kalogera, Vicky. "Three Cosmic Chirps and Counting . . ." *Sky and Telescope* (September 2017): 24–31.

Kaplan, Sarah, and Joel Achenbach. "See a Black Hole for the First Time in a Historic Image from the Event Horizon Telescope." *Washington Post*, April 10, 2019. https://www.washingtonpost.com/science/2019/04/10/see-black-hole-first-time-images-event-horizon-telescope/.

Karr, Barry. "Skeptical Inquirer Presents Series a Big Success: Fills CSICon Gap." *Skeptical Inquirer* 45, no. 3, (May/June 2021).

Kaufman, Allison B., and James C. Kaufman, eds. *Pseudoscience: The Conspiracy against Science*. Cambridge, MA: MIT Press, 2019.

Kemmerer, Stephanie. "Life, the Quniverse, and Everything, Part 1." *Skeptical Inquirer* 45, no. 2 (March/April 2021).

———. "Life, the Quniverse, and Everything, Part 2: QManTrafficking and the 'Plandemic.'" *Skeptical Inquirer* 45, no. 3 (May/June 2021).

Kerr, Richard A. "Mega-Eruptions Drove the Mother of Mass Extinctions." *Science* 342, no. 6165 (December 20, 2013): 1424. https://www.science.org/doi/10.1126/science.342.6165.1424#:~:text=The%20result%3A%20%22We%20can%20say,at%20251.880%20million%20years%20ago.

Kitcher, Philip. *Abusing Science: The Case against Creationism*. Cambridge, MA: MIT Press, 1984.

———. "The MJ-12 Crashed-Saucer Documents." *Skeptical Inquirer* 12, no. 2 (Winter 1987/1988).

———. *UFOs Explained*. New York: Random House, 1974.

Klass, Philip J. *The Real Roswell Crashed-Saucer Coverup*. Amherst, NY: Prometheus Books, 1997.

———. *UFO Abductions: A Dangerous Game*. Updated ed. Amherst, NY: Prometheus Books, 1989.

———. *UFOs: The Public Deceived*. Amherst, NY: Prometheus Books, 1983.

Koertge, Noretta. "Belief Buddies versus Critical Communities." In *Philosophy of Pseudoscience: Reconsideration of the Demarcation Problem*, edited by Massimo Pigliucci and Maarten Boudry. Chicago: University of Chicago Press, 2013.

Kolbert, Elizabeth. "That's What You Think: Is the Appetite for Conspiracy Theories Really New?" *New Yorker*, April 22, 2019, 28–31.

Krebs, Robert E. *Scientific Development and Misconceptions through the Ages: A Reference Guide*. Westport, CT: Greenwood Press, 1999.

Kühlbrandt, Werner. "The Resolution Revolution." *Science* 343, no. 6178 (March 28, 2014): 1443–44. https://www.science.org/doi/10.1126/science.1251652.

Kupferschmidt, Kal. "On the Trail of Bullshit: Studying Misinformation Should Become a Top Scientific Priority, Says Biologist Carl Bergstrom." *Science* 375, no. 6587 (March 25, 2022): 1334–37.

Kurtz, Paul. "Editorial: New Directions for Skeptical Inquiry." *Skeptical Inquirer* 31, no. 1 (January/February 2007): 7.

———. *Exuberant Skepticism.* Edited by John Shook. Amherst, NY: Prometheus Books, 2010.

———. "From the Chairman." *Skeptical Inquirer* 5, no. 3 (Spring 1981): 2–4.

———. *The New Skepticism: Inquiry and Reliable Knowledge.* Amherst, NY: Prometheus Books, 1992.

———. "A Quarter Century of *Skeptical Inquiry*: My Personal Involvement." *Skeptical Inquirer* 25, no. 4 (July/August 2001): 42–47.

———. "Science and the Public: Summing Up Thirty Years of the *Skeptical Inquirer.*" *Skeptical Inquirer* 30, no. 5 (September/October 2006): 13–19.

———, ed. *Skeptical Odysseys: Personal Accounts by the World's Leading Paranormal Inquirers.* Amherst, NY: Prometheus Books, 2001.

———. *A Skeptic's Handbook of Parapsychology.* Amherst, NY: Prometheus Books, 1985.

———. *The Transcendental Temptation: A Critique of Religion and the Paranormal.* Amherst, NY: Prometheus Books, 1991.

Kusche, Lawrence David. *The Bermuda Triangle Mystery—Solved.* New York: Harper and Row, 1975.

Ladendorf, Bob, and Brett Ladendorf. "Wildlife Apocalypse: How Myths and Superstitions Are Driving Animal Extinctions." *Skeptical Inquirer* 42, no. 4 (July/August 2018): 30–39.

Ladyman, James. "Toward a Demarcation of Science from Pseudoscience." In *Philosophy of Pseudoscience: Reconsideration of the Demarcation Problem*, edited by Massimo Pigliucci and Maarten Boudry. Chicago: University of Chicago Press, 2013.

Laudan, Larry. "The Demise of the Demarcation Problem." In *But Is It Science?* edited by Michael Ruse. Amherst, NY: Prometheus Books, 1996.

———. "The Demise of the Demarcation Problem." In *Physics, Philosophy and Psychoanalysis: Essays in Honor of Adolf Grünbaum*, edited by R. S. Cohen and L. Lauden, 111–27. Dordrecht, Holland: D. Reidel, 1983.

Law, Stephen. *Believing Bullshit: How Not to Get Sucked into an Intellectual Black Hole.* Amherst, NY: Prometheus Books, 2011.

Le Page, Michael. "Heatwave in China Is the Most Severe Ever Recorded in the World." *New Scientist*, August 23, 2002. https://www.newscientist

.com/article/2334921-heatwave-in-china-is-the-most-severe-ever-recorded-in-the-world/.

Levinovitz, Alan. "Chairman Mao Invented Traditional Chinese Medicine." Slate. October 22, 2013. https://slate.com/technology/2013/10/traditional-chinese-medicine-origins-mao-invented-it-but-didnt-believe-in-it.html.

Levitan, Dave. *Not a Scientist: How Politicians Mistake, Misrepresent, and Utterly Mangle Science.* New York: W. W. Norton, 2017.

Lewens, Tim. *The Meaning of Science.* London: Penguin Random House UK, 2015.

Lilienfeld, Scott O. "Foreword: Navigating a Post-truth World: Ten Enduring Lessons from the Study of Pseudoscience." In *Pseudoscience: The Conspiracy against Science*, edited by Allison B. Kaufman and James C. Kaufman. Cambridge, MA: MIT Press, 2018.

———. "Intellectual Humility: A Guiding Principle for the Skeptical Movement?" *Skeptical Inquirer* 44, no. 5 (September/October 2020): 32–37.

Lilienfeld, Scott O., John Ruscio, and Steven Jay Lynn, eds. *Navigating the Mindfield: A Guide to Separating Science from Pseudoscience in Mental Health.* Amherst, NY: Prometheus Books, 2008.

Lindsey, Rebecca. "Climate Change: Atmospheric Carbon Dioxide." National Oceanic and Atmospheric Administration. June 23, 2022. https://www.climate.gov/news-features/understanding-climate/climate-change-atmospheric-carbon-dioxide.

Loxton, Daniel. "Why Is There a Skeptical Movement? Part One: Two Millennia of Paranormal Skepticism." 2013. https://www.arvindguptatoys.com/arvindgupta/skeptical-movement.pdf.

Lumiere, Frederic, dir. *Lady Ganga: Nilza's Story.* Lumiere Media, 2015.

Mackay, Charles. *Extraordinary Popular Delusions and the Madness of Crowds.* New York: Harmony Books, 1980.

"Magicians, Skeptics Share Their Memories of James Randi." *Skeptical Inquirer* 45, no. 1 (January/February 2021): 44.

Mahner, Martin. "Science and Pseudoscience." In *Philosophy of Pseudoscience: Reconsideration of the Demarcation Problem*, edited by Massimo Pigliucci and Maarten Boudry. Chicago: University of Chicago Press, 2013.

Mangold, N., S. Gupta, O. Gasnault, G. Dromart, J. D. Tarnas, S. F. Sholes, B. Horgan, et al. "Perseverance Rover Reveals an Ancient Delta-Lake System and Flood Deposits at Jezero Crater, Mars." *Science* 374, no. 6568 (October 7, 2021): 711–17.

Mann, Michael E. *The Hockey Stick and the Climate Wars: Dispatches from the Front Lines.* New York: Columbia University Press, 2012.

———. "How to Win the New Climate War." *Skeptical Inquirer* 44, no. 2 (March/April 2020): 18–19.

Mann, Michael E., and Lee R. Kump. *Dire Predictions: Understanding Global Warming*. London: Pearson Education, 2009.

Mann, Michael E., and Lee Toles. *The Madhouse Effect: How Climate Change Denial Is Threatening Our Planet, Destroying Our Politics, and Driving Us Crazy*. 2016. New York: Columbia University Press, 2016.

Masson-Delmotte, Valérie, Panmao Zhai, Hans-Otto Pörtner, Debra Roberts, Jim Skea, Priyadarshi R. Shukla, Anna Pirani, Wilfran Moufouma-Okia, Clotilde Péan, Roz Pidcock, et al. *Global Warming of 1.5°C: An IPCC Special Report on the Impacts of Global Warming of 1.5°C above Pre-industrial Levels and Related Global Greenhouse Gas Emission Pathways, in the Context of Strengthening the Global Response to the Threat of Climate Change, Sustainable Development, and Efforts to Eradicate Poverty*. Cambridge, UK: Cambridge University Press, 2019.

Matthews, Michael R., ed. *Mario Bunge: A Centenary Festschrift*. Cham, Switzerland: Springer, 2019.

———. "Mario Bunge: Physicist, Philosopher, Champion of Science, Citizen of the World." *Skeptical Inquirer* 44, no. 4 (July/August 2020): 7–8.

Mazziotta, Julie. "Goop to Pay $145,000 Settlement for Misleading Claims about the Effectiveness of Vaginal Eggs." *People*, September 5, 2018. https://people.com/health/gwyneth-paltrow-goop-pay-settlement-misleading-claims-vaginal-eggs/.

McIntyre, Lee. *The Scientific Attitude: Defending Science from Denial, Fraud, and Pseudoscience*. Cambridge, MA: MIT Press, 2019.

Medawar, Peter. "Two Conceptions of Science." In *Pluto's Republic*. Oxford, UK: Oxford University Press, 1984.

Menzel, Donald H. *Flying Saucers*. Cambridge, MA: Harvard University Press, 1953.

Menzel, Donald H., and Ernest H. Taves. *The UFO Enigma: The Definite Explanation of the UFO Phenomenon*. Garden City, NY: Doubleday, 1977.

Montes de Oca, Bernardo. "What Is Goop? Here's Why Everyone Hates Gwyneth Paltrow's Company." Slidebean. September 2, 2021. https://slidebean.com/story/what-is-goop-paltrow-slidebean#:~:text=Goop%20claimed%20that%20its%20Body,said%20it%20was%20an%20error.

Morrison, David. "Velikovsky at Fifty: Cultures in Collision on the Fringes of Science." *Skeptic* 9, no. 1 (2021).

National Academy of Sciences Institute of Medicine. *Science, Evolution, and Creationism*. Washington, DC: National Academies Press, 2008.

National Aeronautics and Space Administration. "Graphic: The Relentless Rise of Carbon Dioxide." Accessed April 15, 2023. https://climate.nasa.gov/climate_resources/24/graphic-the-relentless-rise-of-carbon-dioxide/.

———. "NASA Telescope Reveals Largest Batch of Earth-Size, Habitable-Zone Planets around Single Star." Release 17-015, February 22, 2017. https://www.nasa.gov/press-release/nasa-telescope-reveals-largest-batch-of-earth-size-habitable-zone-planets-around.

———. *Webb Telescope: Media Kit.* Washington, DC: NASA, 2021. https://www.webb.nasa.gov/content/webbLaunch/assets/documents/WebbMediaKit.pdf.

National Aeronautics and Space Administration, Goddard, and Arizona State University. "NASA Releases New High-Resolution Earthrise Image." December 18, 2015. http://www.nasa.gov/image-feature/goddard/lro-earthrise-2015.

National Aeronautics and Space Administration, JPL-Caltech, and Space Science Institute. "The Day the Earth Smiled." November 12, 2013. https://saturn.jpl.nasa.gov/resources/5916/.

———. "One Special Day in the Life of Planet Earth." July 22, 2013. https://saturn.jpl.nasa.gov/resources/5864/.

National Oceanic and Atmospheric Administration. "Carbon Dioxide Now More Than 50% Higher Than Pre-industrial Levels." June 3, 2022. https://www.noaa.gov/news-release/carbon-dioxide-now-more-than-50-higher-than-pre-industrial-levels.

National Science Board. "Science and Technology: Public Attitudes and Understanding." In *Science Indicators 2014.* Arlington, VA: National Science Foundation, 2014.

Newburger, Emma. "UN Warns World Is Entering 'Uncharted Territories of Destruction' from Climate Crisis." CNBC. September 13, 2022. https://www.cnbc.com/2022/09/13/world-entering-uncharted-territories-of-destruction-climate-crisis-un.html#:~:text=The%20United%20Nations%20is%20warning,to%20reduce%20greenhouse%20gas%20emissions.

Newman, Steve. "Earthweek: Diary of a Changing World: Week Ending August 12, 2022." https://www.earthweek.com/_files/ugd/532fc1_2ccad38118d442fd8e5900ce92385aea.pdf.

———. "Earthweek: Diary of a Changing World: Week Ending, September 16, 2022." https://www.earthweek.com/_files/ugd/532fc1_30e9825882994c87aee3feeef999c3b0.pdf.

Nickell, Joe. "'Alien Autopsy' Hoax." *Skeptical Inquirer* 19, no. 6 (November/December 1995).

———. *Inquest on the Shroud of Turin.* Amherst, NY: Prometheus Books, 1983.

———. "Premonition! Foreseeing What Cannot Be Seen." *Skeptical Inquirer* 43, no. 4 (July/August 201o9): 17–20.

———. *The Science of Miracles: Investigating the Incredible.* Amherst, NY: Prometheus Books, 2013.

Nicolaides, Demetris. *In the Light of Science: Our Ancient Quest for Knowledge and the Measure of Modern Physics.* Amherst, NY: Prometheus Books, 2014.

Nicolotti, Andrea. *The Shroud of Turin: The History and Legends of the World's Most Famous Relic.* Translated by Jeffrey M. Hunt and R. A. Smith. Waco, TX: Baylor University Press, 2019.

Novella, Steven, with Bob Novella, Cara Santa Maria, Jay Novella, and Evan Bernstein. *The Skeptics' Guide to the Universe: How to Know What's Really Real in a World Increasingly Full of Fake.* New York: Grand Central, 2018.

Nurk, Sergey, Sergey Koren, Arang Rhie, Mikko Rautiainen, Andrey V. Bzikadze, Alla Mikheenko, Mitchell R. Vollger, Nicolas Altemose, Lev Uralsky, Ariel Gershman, et al. "The Complete Sequence of the Human Genome." *Science* 376, no. 6588 (April 1, 2022):44–53. https://www.science.org/doi/10.1126/science.abj6987.

Offit, Paul A. *Bad Advice: Or Why Celebrities, Politicians, and Activists Aren't Your Best Source of Health Information.* New York: Columbia University Press, 2018.

———. *Deadly Choices: How the Anti-Vaccine Movement Threatens Us All.* New York: Basic Books, 2015.

———. *Do You Believe in Magic? The Sense and Nonsense of Alternative Medicine.* New York: HarperCollins, 2013.

Oreskes, Naomi. "Can Science Be Saved?" *Skeptical Inquirer Presents* (webinar). October 28, 2021. https://skepticalinquirer.org/video/can-science-be-saved-with-naomi-oreskes/.

Oreskes, Naomi, and Erik M. Conway. *Merchants of Doubt: How a Handful of Scientists Obscured the Truth on Issues from Tobacco Smoke to Global Warming.* New York: Bloomsbury Press, 2010.

Oswald, Mark. "Shirley MacLaine Selling Abiquiu Ranch." *Albuquerque Journal*, April 19, 2014, C1. https://www.abqjournal.com/386415/maclaine-selling-ranch.html.

Otto, Shawn. *The War on Science: Who's Waging It, Why It Matters, and What We Can Do about It.* Minneapolis: Milkweed Editions, 2016.

Overbye, Dennis. "Darkness Visible, Finally: Astronomers Capture First Ever Image of a Black Hole." *New York Times*, April 10, 2019. https://www.nytimes.com/2019/04/10/science/black-hole-picture.html.

Pankratz, Loren. *Mysteries and Secrets Revealed: From Oracles at Delphi to Spiritualism in America.* Lanham, MD: Prometheus Books, 2021.

Park, Robert. *Superstition: Belief in the Age of Science*. Princeton, NJ: Princeton University Press, 2008.

———. *Voodoo Science: The Road from Foolishness to Fraud*. Oxford, UK: Oxford University Press, 2000.

Parson, Keith. *It Started with Copernicus: Vital Questions about Science*. Amherst, NY: Prometheus Books, 2014.

———, ed. *The Science Wars: Debating Scientific Knowledge and Technology*. Amherst, NY: Prometheus Books, 2003.

Peebles, Curtis. *Watch the Skies! A Chronicle of the Flying Saucer Myth*. Washington, DC: Smithsonian Institution Press, 1994.

Pennisi, Elizabeth. "The Power of Many." *Science* 360, no. 6396 (June 29, 2018): 1388. https://www.science.org/doi/abs/10.1126/science.360.6396.1388.

Petrocelli, John V. *The Life-Changing Science of Detecting Bullshit*. New York: St. Martin's Press, 2021.

Picin, Andrea, Stefano Benazzi, Ruth Blasco, Mateja Hajdinjak, Kristofer M. Helgen, Jean-Jacques Hublin, Jordi Rosell, Pontus Skoglund, Chris Stringer, and Sahra Talamo. "Comment on 'A Global Environmental Crisis 42,000 Years Ago." *Science* 374, no. 6570 (November 18, 2021). https://doi.org/10.1126%2Fscience.abi8330.

Pigliucci, Massimo. "The Borderlands between Science and Philosophy: An Introduction." *Quarterly Review of Biology* 83, no. 1 (March 2008): 7–15.

———. "The Demarcation Problem: A (Belated) Response to Laudan." In *Philosophy of Pseudoscience: Reconsideration of the Demarcation Problem*, edited by Massimo Pigliucci and Maarten Boudry. Chicago: University of Chicago Press, 2013.

———. *Nonsense on Stilts: How to Tell Science from Bunk*. 2nd ed. Chicago: University of Chicago Press, 2018.

———. "Science Denialism Is a Form of Pseudoscience." *Skeptical Inquirer* 46, no. 1 (January/February 2022): 19–20.

Pigliucci, Massimo, and Maarten Boudry. "The Dangers of Pseudoscience." *The Stone, New York Times Opinionator* (blog). October 10, 2013. http://opinionator.blogs.nytimes.com/2013/10/10/the-dangers-of-pseudoscience/.

———, eds. *Philosophy of Pseudoscience: Reconsideration of the Demarcation Problem*. Chicago: University of Chicago Press, 2013.

Pinker, Steven. *Enlightenment Now: The Case for Reason, Science, Humanism, and Progress*. New York: Viking, 2018.

———. "Enlightenment Wars: Some Reflections on 'Enlightenment Now,' One Year Later." *Quillette*, January 14, 2019. https://quillette.com/2019/01/14/enlightenment-wars-some-reflections-on-enlightenment-now-one-year-later/.

———. "Progressophobia: Why Things Are Better Than You Think They Are." *Skeptical Inquirer* 43, no. 3 (May/June 2018): 26–35.

———. *Rationality: What It Is, Why It Seems Scarce, Why It Matters.* New York: Viking, 2021.

Plait, Philip. *Bad Astronomy: Misconception and Misuses Revealed, from Astrology the Moon Landing "Hoax."* New York: Wiley, 2002.

Planck Collaboration. "*Planck* Intermediate Results: XXX. The Angular Power Spectrum of Polarized Dust Emission at Intermediate and High Galactic Latitudes." *Astronomy and Astrophysics* 586 (February 2016). https://doi.org/10.1051/0004-6361/201425034.

Planck, Max. *Scientific Biography and Other Papers: With a Memorial Address on Max Planck.* New York: Philosophical Library, 1949.

Polidoro, Massimo. "Stop the Epidemic of Lies: Thinking about COVID-19 Misinformation." *Skeptical Inquirer* 44, no. 4 (July/August 2020): 15–16.

Pollack, Henry. Lecture at a meeting of New Mexicans for Science and Reason, December 2, 2009.

———. *A World without Ice.* New York: Avery/Penguin Group, 2009.

Pollan, Michael. "The Intelligent Plant." *New Yorker*, December 23 and 30, 2013, 92.

Pörtner, Hans-Otto, Debra C. Roberts, Melinda M. B. Tignor, Elvira Poloczanska, Katja Mintenbeck, Andrés Alegría, Marlies Craig, et al., eds. *Climate Change 2022: Impacts, Adaptation and Vulnerability: Working Group II Contribution to the Sixth Assessment Report of the Intergovernmental Panel on Climate Change.* Cambridge, UK: Cambridge University Press, 2022.

Powers, Richard. *The Overstory.* New York: W. W. Norton, 2018.

Prothero, Donald R., and Timothy D. Callahan. *UFOs, Chemtrails, and Aliens: What Science Says.* Bloomington: Indiana University Press, 2018.

Quantum Analyzer. "What Is Quantum Analyzer AH-Q8? Quantum Resonance Magnetic Analyzer." 2018. http://www.quantum-resonance-magnetic-analyzer.com/quantum-analyzer-ah-q8.html.

Radford, Benjamin. *America the Fearful: Media and the Marketing of National Panics.* Jefferson, NC: McFarland, 2022.

———. *Big If True: Adventures in Oddity.* Corrales, NM: Rhombus Books, 2021.

———. "Coronavirus Crisis: Chaos, Counting, and Confronting Our Biases." *Skeptical Inquirer* 44, no. 4 (July/August 2020): 10–14.

———. "Intuition Nearly Kills Hiker." *A Skeptic Reads the Newspaper* (blog). Center for Inquiry. July 19, 2019. https://centerforinquiry.org/blog/intuition-nearly-kills-hiker/.

———. "Jan. 6 Investigation Testimony Reveals Conspiracy-Riddled Trump White House." *Skeptical Inquirer* 44, no. 5 (September/October 2022): 5–6.

———. *Mysterious New Mexico: Miracles, Magic, and Monsters in the Land of Enchantment.* Albuquerque: University of New Mexico Press, 2014.

———. "Psychic Defective: Sylvia Browne Blunders Again." *Skeptical Inquirer* 37, no. 3 (July/August 2013): 7–8.

———. *Scientific Paranormal Investigation: How to Solve Unexplained Mysteries.* Corrales, NM: Rhombus Books, 2010.

Randi, James. *Conjuring*. New York: St. Martin's Press, 1992.

———. *An Encyclopedia of Claims, Frauds, and Hoaxes of the Occult and Supernatural.* New York: St. Martin's Press, 1995.

———. *Flim-Flam! Psychics, ESP, Unicorns, and Other Delusions*. Lanham, MD: Prometheus Books, 2022.

———. *The Magic of Uri Geller*. New York: Ballantine Books, 1975.

Read, Piers Paul. *Alive: The Story of the Andes Survivors*. Philadelphia: Lippincott, 1974.

Reber, Arthur S., and James E. Alcock. "Searching for the Impossible: Parapsychology's Elusive Quest." *American Psychologist* 75 (2020): 391–99.

———. "Why Parapsychological Claims Cannot Be True." *Skeptical Inquirer* 43, no. 4 (July/August 2019): 8–10. https://skepticalinquirer.org/2019/07/why-parapsychological-claims-cannot-be-true/.

Revelle, Roger. "Carbon Dioxide and World Climate." *Scientific American* 247, no. 2 (August 1982).

Revelle, Roger, and Hans E. Suess. "Carbon Dioxide Exchange between Atmosphere and Ocean and the Question of an Increase of Atmospheric CO_2 during the Past Decades." *Tellus* 9, no. 1 (February 1957): 18–27.

Richards, Dana, ed. *Dear Martin, Dear Marcello: Gardner and Truzzi on Skepticism.* Hackensack, NJ: World Scientific, 2017.

Richer, Alanna Durkin. "Explainer: Hundreds Charged with Crimes in Capitol Attack." Associated Press. June 7, 2022. https://apnews.com/article/capitol-siege-merrick-garland-government-and-politics-conspiracy-crime-c2e427dc0fa16077d7fb98c06e61149f.

Roberts, Walter Orr, and Henry Lansford. *The Climate Mandate*. San Francisco: W. H. Freeman, 1979.

Rood, Jennifer E., and Aviv Regev. "The Legacy of the Human Genome Project." *Science* 373, no. 6562 (September 24, 2021): 1442–43. https://www.science.org/doi/10.1126/science.abl5403.

Rosling, Hans, and Ola Rosling. "A Fact-Based Worldview through Animated Data." Presentation at "ESCape to Clarity!" Fifteenth European Skeptics Congress, Stockholm, Sweden, August 23–25, 2013.

Rosling, Hans, with Ola Rosling and Anna Rosling Rönnlund. *Factfulness: Ten Reasons We're Wrong about the World—and Why Things Are Better Than You Think.* New York: Flatiron Books, 2018.

Rothman, Joshua. "The Big Question: Is the World Getting Better or Worse?" *New Yorker*, July 28, 2018, 26–32.

Rothschild, Mike. *The Storm Is upon Us: How QAnon Became a Movement, Cult, and Conspiracy Theory of Everything*. Brooklyn, NY: Melville House, 2021.

Ruse, Michael. *But Is It Science?* Amherst, NY: Prometheus Books, 1996.

———. "Democracy and the Problem of Pseudo-science." In *Anti-Science and the Assault on Democracy*, edited by Michael J. Thompson and Gregory Smulewicz-Zucker. Amherst, NY: Prometheus Books, 2018.

Sagan, Carl. *Broca's Brain: Reflections on the Romance of Science.* New York: Random House, 1979.

———. "The Burden of Skepticism." *Skeptical Inquirer* 12, no. 1 (Fall 1987): 38–46.

———. *The Demon-Haunted World.* New York: Random House, 1995.

———. "Night Walkers and Mystery Mongers: Sense and Nonsense at the Edge of Science." *Skeptical Inquirer* 10, no. 3 (Spring 1986): 218–28.

———. "Wonder and Skepticism." *Skeptical Inquirer* 19, no. 1 (January/February 1995): 24–30.

Sandefur, Timothy. *The Ascent of Jacob Bronowski: The Life and Ideas of a Popular Science Icon.* Amherst, NY: Prometheus Books, 2019.

Schick, Theodore, Jr., and Lewis Vaughn. *How to Think about Weird Things: Critical Thinking for New Age.* Mountain View, CA: Mayfield, 1995.

Schlagel, Richard H. *Seeking the Truth: How Science Has Prevailed over the Supernatural Worldview.* Amherst, NY: Humanity Books, 2011.

Schmidt, Amanda. "UN Climate Change Panel Says 'Unprecedented Changes' Needed to Prevent Rapid Global Warming." Yahoo! News. October 8, 2018. https://www.yahoo.com/news/un-climate-change-panel-says-173744678.html.

Schneider, Stephen H. "Climate Change: Skeptics vs. Deniers." *Skeptical Inquirer* 33, no. 3 (May/June 2009): 6.

———. *Global Warming: Are We Entering the Greenhouse Century?* San Francisco: Sierra Club Books, 1989.

———. *Science as a Contact Sport: Inside the Battle to Save Earth's Climate.* Washington, DC: National Geographic, 2009.

Schneider, Stephen H., and Randi Londer. *The Coevolution of Climate and Life.* San Francisco: Sierra Club Books, 1984.

Schneider, Stephen H., with Lynne E. Mesirow. *The Genesis Strategy: Climate and Global Survival.* New York: Plenum Press, 1976.

Select Committee on Aging. *Quackery: A $10 Billion Scandal.* H.R. Doc. 98-262 (1984).

Sense about Science. "Maddox Prize 2015." Accessed April 11, 2023. https://senseaboutscience.org/activities/2015-john-maddox-prize/.

Service, Robert F. "Protein Structures for All." *Science* 374, no. 6574 (December 17, 2021): 1426–28.

Shaffer, Ryan. "The Psychic Defective Revisited." *Skeptical Inquirer* 37, no. 5 (September/October 2013): 30–35. http://www.csicop.org/si/show/the_psychic_defective_revisited_years_later_sylvia_brownes_accuracy_remains/.

Shaffer, Ryan, and Agatha Jadwiszczok. "Psychic Defective: Sylvia Browne's History of Failure." *Skeptical Inquirer* 34, no. 2 (March/April 2010): 38–42. https://skepticalinquirer.org/2020/03/psychic-defective-sylvia-brownes-history-of-failure/.

Sheaffer, Robert. "Aztec Saucer Crash Story Rises from the Dead." *Skeptical Inquirer* 36, no. 6 (November/December 2012).

———. *Bad UFOs: Skepticism, UFOs, and the Universe* (blog). Accessed April 16, 2023. www.badUFOs.com.

———. *Psychic Vibrations: Skeptical Giggles from the* Skeptical Inquirer. Illustrations by Rob Pudim. Charleston, SC: Create Space, 2011.

———. *The UFO Verdict: Examining the Evidence.* Amherst, NY: Prometheus Books, 1981.

Sheldrake, Merlin. *Entangled Life: How Fungi Make Our Worlds, Change Our Minds, and Shape Our Futures*. New York: Random House, 2020.

Shermer, Michael. *The Borderlands of Science: Where Sense Meets Nonsense.* Oxford, UK: Oxford University Press, 2001.

———. *Science Friction: Where the Known Meets the Unknown.* New York: Times Books, 2005.

Sidky, H. *Religion, Supernaturalism, the Paranormal, and Pseudoscience*. New York: Anthem Press, 2020.

———. *Science and Anthropology in a Post-Truth World: A Critique of Unreason and Academic Nonsense.* Lanham, MD: Rowman and Littlefield, 2021.

———. "The War on Science and Knowledge: Anti-Intellectualism in Today's America." *Skeptical Inquirer* 42, no. 2 (March/April 2018): 38–43.

Simard, Suzanne. *Finding the Mother Tree: Discovering the Wisdom of the Forest.* New York: Alfred A. Knopf, 2021.

Sinatra, Gale M., and Barbara K. Hofer. *Science Denial: Why It Happens and What to Do about It.* Oxford, UK: Oxford University Press, 2021.

Singh, Simon, and Edzard Ernst. *Trick or Treatment? The Undeniable Facts about Alternative Medicine.* New York: W. W. Norton, 2008.

The Skeptic. "About 'The Skeptic.'" Accessed April 16, 2023. https://www.skeptic.org.uk/about/.

Small-M. "Math! How Much CO_2 by Weight in the Atmosphere?" March 30, 2007. https://micpohling.wordpress.com/2007/03/30/math-how-much-co2-by-weight-in-the-atmosphere/.

Smith, Jonathan C. *Pseudoscience and Extraordinary Claims of the Paranormal: A Critical Thinker's Toolkit.* Malden, MA: Wiley-Blackwell, 2010.

Smith, Tracy. "In Conversation: Gwyneth Paltrow." *CBS News Sunday Morning.* Aired September 25, 2022.

Snellen, Ignas A. G. "Astronomy: Earth's Seven Sisters." *Nature* 542 (February 23, 2017): 421–23.

Social Media Storm. Special issue. *Science* 375, no. 6587 (March 25, 2022): 1332–47.

Sparks, Matthew. "DeepMind's Protein-Folding AI Cracks Biology's Biggest Problem." *New Scientist*, July 28, 2022. https://www.newscientist.com/article/2330866-deepminds-protein-folding-ai-cracks-biologys-biggest-problem/.

Spinoza, Baruch. *Tractatus Politicus* (1677). In *The Oxford Dictionary of Quotations*, 3rd ed. Oxford, UK: Oxford University Press, 1979.

Stephens-Davidowitz, Seth. *Everybody Lies: Big Data, New Data, and What the Internet Can Tell Us about Who We Really Are.* New York: HarperCollins, 2017.

Stokes, Trey. "How to Make an 'Alien' for 'Autopsy.'" *Skeptical Inquirer* 20, no. 1 (January/February 1996).

Strahler, Arthur N. *Understanding Science: An Introduction to Concepts and Issues.* Amherst, NY: Prometheus Books, 1992.

Sullivan, Martin J. P., Simon L. Lewis, Kofi Affum-Baffoe, Carolina Castilho, Flávia Costa, Aida Cuni Sanchez, Corneille E. N. Ewango, et al. 2020. "Long-Term Thermal Sensitivity of Earth's Tropical Forests." *Science* 368, no. 6493 (May 22, 2020): 869–74.

Tavris, Carol, and Elliot Aronson. *Mistakes Were Made (but Not by Me): Why We Justify Foolish Beliefs, Bad Decisions, and Hurtful Acts.* Updated ed. New York: Harcourt, 2020.

Tchekmedyian, Alene. "Gwyneth Paltrow's Goop to Offer Refunds over 'Unsubstantiated' Claims about Health Benefits." *Los Angeles Times*, September 4, 2018. https://www.latimes.com/local/lanow/la-me-ln-goop-settlement-20180904-story.html.

Temming, Maria. "100 Years of *Science News*." *Science News* 201, no. 6 (March 26, 2022): 16–25.

Thacker, Paul. "Fred Singer Has Passed. May He Rest." Drilled News, April 16, 2020. www.drillednews.com/post/fred-singer-obituary-climate-denier.

Thomas, David E. "The Aztec UFO Symposium: How This Saucer Story Started as a Con Game." *Skeptical Inquirer* 22, no.5 (September/October 1998).

———. "The 'Roswell Fragment'—Case Closed." *Skeptical Inquirer* 20, no. 6 (November/December 1996).

Thompson, Michael J., and Gregory R. Smulewicz-Zucker. *Anti-Science and the Assault on Democracy: Defending Reason in a Free Society.* Amherst, NY: Prometheus Books, 2018.

Thorp, H. Holden. "Editorial: Proteins, Proteins Everywhere." *Science* 374, no. 6574 (December 17, 2021): 1415. https://www.science.org/doi/10.1126/science.abn5795.

———. "Letters to the Editor: The State of Our Nation." *Skeptical Inquirer* 45, no. 1 (January/February 2021): 64.

Thurs, Daniel P., and Ronald L. Numbers. "Science, Pseudoscience, and Science Falsely So-Called." In *Philosophy of Pseudoscience: Reconsideration of the Demarcation Problem*, edited by Massimo Pigliucci and Maarten Boudry, 121–44. Chicago: University of Chicago Press, 2013.

Tiller, Nick. "From Debunking to Prebunking: How Skeptical Activism Must Evolve to Meet the Growing Anti-science Threat." *Skeptical Inquirer* 46, no. 5 (September/October 2022).

Trecek-King, Melanie. "How to Sell Pseudoscience." *Skeptical Inquirer* 46, no. 5 (September/October 2022): 46–49.

———. "A Life Preserver for Staying Afloat in a Sea of Misinformation." *Skeptical Inquirer* 46, no. 2 (March/April 2022): 44–49.

———. "Teach Skills, Not Facts." *Skeptical Inquirer* 46, no. 1 (January/February 2022): 34–38.

Truzzi, Marcello. "Pseudoscience." In *The Encyclopedia of the Paranormal*, edited by Gordon Stein, 560–75. Amherst, NY: Prometheus Books, 1996.

UC Museum of Paleontology. "Prepare and Plan: Correcting Misconceptions." University of California, Berkeley. 2023. http://undsci.berkeley.edu/teaching/misconceptions.php#b1.

UK Web Archive. Accessed April 16, 2023. https://www.webarchive.org.uk.

Uscinski, Joseph E. "Clear Thinking about Conspiracy Theories in These Troubled Times." *Skeptical Inquirer* 45, no. 1 (January/February 2021): 52–56.

———. "Conspiring for the Common Good." *Skeptical Inquirer* 43, no. 4 (July/August 2019): 40–44.

Uscinski, Joseph E., and Joseph M. Parent. *American Conspiracy Theories.* New York: Oxford University Press, 2004.

Varadi, Mihaly, Stephen Anyango, Mandar Deshpande, Sreenath Nair, Cindy Natassia, Galabina Yordanova, David Yuan, Oana Stroe, Gemma Wood, Agata Laydon, et al. "AlphaFold Protein Structure Database: Massively Expanding the Structural Coverage of Protein-Sequence Space with High-Accuracy Models." *Nucleic Acids Research* 50 (January 1, 2022): D439–44. https://doi.org/10.1093/nar/gkab1061.

"Velikovsky: The Controversy Continues." *New York Times Book Review*, April 20, 1980.

Velikovsky, Immanuel. *Worlds in Collision*. New York: Macmillan, 1950.

Verschuuren, Gerard M. *Life Scientists: Their Convictions, Their Activities, Their Values*. North Andover, MA: Genesis, 1995.

Victor, Jeffrey S. *Satanic Panic: The Creation of a Contemporary Legend*. Chicago: Open Court, 1993.

———. "The Social Dynamics of Conspiracy Rumors: From Satanic Panic to QAnon." *Skeptical Inquirer* 46, no. 4 (July/August 2022): 37–41.

———. "The Social Forces behind Trump's War on Science." *Free Inquiry* 38, no. 5 (August/September 2018).

Vitale, Salvatore. "The First 5 Years of Gravitational-Wave Astrophysics." *Science* 372, no. 6546 (June 4, 2021): 1054. https://www.science.org/doi/10.1126/science.abc7397.

Voosen, Paul. "Kauri Trees Mark Magnetic Flip 42,000 Years Ago." *Science* 371, no. 6531 (February 19, 2021): 766. https://www.science.org/doi/10.1126/science.371.6531.766.

Vyse, Stuart. *Believing in Magic: The Psychology of Superstition*. Updated ed. Oxford, UK: Oxford University Press, 2013.

———. "Why Your Uncle Isn't Going to Get Vaccinated." *Skeptical Inquirer* 46, no. 10 (January/February 2022): 24–29.

Wainer, Howard. *Truth or Truthiness: Distinguishing Fact from Fiction by Learning to Think like a Data Scientist*. New York: Cambridge University Press, 2016.

Washington, Haydn, and John Cook. *Climate Change Denial: Heads in the Sand*. London: Earthscan.

Weinberg, Steven. *Facing Up: Science and Its Cultural Adversaries*. Cambridge, MA: Harvard University Press, 2001.

———. *To Explain the World: The Discovery of Modern Science*. New York: HarperCollins, 2015.

Weisskopf, Victor F. *Knowledge and Wonder*. New York: Anchor Books, 1963.

Welch, Ethan L. *Quackonomics: The Cost of Unscientific Healthcare in the United States*. Conneaut Lake, PA: Page, 2020.

West, Jevin D., and Carl. T. Bergstrom. "Misinformation in and about Science." *Proceedings of the National Academy of Sciences* 118, no. 15 (April 9, 2021).

West, Mick. *Escaping the Rabbit Hole: How to Debunk Conspiracy Theories Using Facts, Logic, and Respect*. New York: Skyhorse, 2018.

———. "Great Expectations: House Hearing on UAP Excites Fans, Offers Little Else." *Skeptical Inquirer* 46, no. 5 (September/October 2022).

Wiseman, Richard. *Paranormality: Why We See What Isn't There.* London: Macmillan, 2011.

Witkowski, Tomasz. *Psychology Led Astray: Cargo Cult in Science and Therapy.* Boca Raton, FL: BrownWalker Press, 2016.

Witkowski, Tomasz, and Maciej Zatonski. *Psychology Gone Wrong: The Dark Sides of Science and Therapy.* Boca Raton, FL: BrownWalker Press, 2015.

Wohlleben, Peter. *The Hidden Life of Trees: What They Feel, How They Communicate: Discoveries from a Secret World.* Translated by Jane Billinghurst. Vancouver, BC: David Suzuki Institute, 2016.

Wolpert, Lewis. *The Unnatural Nature of Science: Why Science Does Not Make (Common) Sense.* Cambridge, MA: Harvard University Press, 1993.

Wrobel, Arthur, ed. *Pseudo-Science and Society in Nineteenth-Century America.* Lexington: University Press of Kentucky, 1987.

Wynn, Charles M., and Arthur W. Wiggins. *Quantum Leaps in the Wrong Direction: Where Real Science Ends—and Pseudoscience Begins.* 2nd ed. With Cartoons by Sydney Harris. New York: Oxford University Press, 2017.

Zimmer, Marc. *Science and the Skeptic: Discerning Facts from Fiction.* Minneapolis: Twenty-First Century Books, 2022.

Zimring, James C. *What Science Is and How It Really Works.* Cambridge, UK: Cambridge University Press, 2019.

INDEX